PrL   C 12 /D

Coding Places

Acting with Technology
Bonnie Nardi, Victor Kaptelinin, and Kirsten Foot, editors

Coding Places

Software Practice in a South American City

Yuri Takhteyev

The MIT Press
Cambridge, Massachusetts
London, England

MIT Press books may be purchased at special quantity discounts for business or sales promotional use. For information, please email special_sales@mitpress.mit.edu or write to Special Sales Department, The MIT Press, 55 Hayward Street, Cambridge, MA 02142.

This book was set in Stone Sans and Stone Serif by the MIT Press. Printed and bound in the United States of America.

Library of Congress Cataloging-in-Publication Data

Takhteyev, Yuri, 1976–
Coding places : software practice in a South American city / Yuri Takhteyev.
    p. cm. — (Acting with technology)
Includes bibliographical references and index.
ISBN 978-0-262-01807-4 (hardcover : alk. paper)
1. Computer software—Development—Brazil. 2. Lua (Computer program language)
3. Computer programming—Brazil. 4. Globalization. I. Title.
QA76.76.D47T345 2012
005.100981—dc23
2012007128

10 9 8 7 6 5 4 3 2 1

To the memory of Vladimir Takhteyev (1953–2011)
and to Dimitri (b. 2012)

Contents

Acknowledgments

This book could not have happened without the many people who have in various forms provided support, guidance, and inspiration.

The book is based primarily on research done while pursuing a doctoral program at the School of Information at the University of California, Berkeley, and I am highly grateful to my many mentors there. From among those, Paul Duguid's role stands out especially. Paul's *Social Life of Information* introduced me to many of the questions that I explore in this book even before I formulated my graduate school plans. Paul's arrival at the School of Information in 2004 was a true blessing. I feel especially thankful to Paul for his willingness to not just share his insights and knowledge, but also to invest so much time in half-incoherent drafts, helping me formulate my thoughts. I am also thankful to Peter Lyman, under whose supervision I started my program at Berkeley. I am grateful to Peter for convincing me to come to Berkeley, for introducing me to ethnography and social theory, and for securing funding for my research. He is and will be greatly missed.

I am also tremendously indebted to other mentors who have inspired me with their own work and have put much time into helping me improve mine. Anno Saxenian's work on transnational connections set an example for me from my early days at Berkeley and encouraged me to do international research. Anno's advice over the years has also been invaluable. Peter Evans's work introduced me to the history of Brazilian IT policy, and his continuous insistence on hearing "the point" of my work helped me sharpen my arguments. I also want to thank Coye Cheshire, Jean Lave, Suzanne Scotchmer, Michael Buckland, Nancy Van House, Ray Larson, Ted Egan, and Raka Ray for introducing me to many new ideas and providing suggestions for my work.

Fellow students at Berkeley were also a source of inspiration and ideas. The diversity of their interests introduced me to a variety of ways of thinking about information technology, compensating for the narrow specialization inherent in a doctoral program. Jens Grossklags, Paul Laskowski, Joseph Hall, danah boyd, and Mahad Ibrahim helped me stay on track; they were a great cohort. I thank Dan Perkel, Megan Finn, Ryan Shaw, Christo Sims, Rajesh Veeraraghavan, and Bob Bell for attending numerous practice talks, providing suggestions, and being a great group to be around. Aaron Shaw and Michael Donovan, who were themselves working on Brazil, were a great source of insights about that country.

Over one hundred people have generously volunteered their time to tell me about their lives and work. This book would not be possible without their courage to share their stories with a stranger. Many of those people also helped me feel at home in a new city. Some have become friends. Many of my interviewees have volunteered to read chapters of this book, finding factual errors, misinterpretations, inconsistencies, and often simply grammatical mistakes. I am particularly grateful to "Rodrigo" for his willingness to talk openly about the many challenges facing his project, for introducing me to many of the people on whose stories this book relies, and for tolerating my straddling of fieldwork and friendship. I thank Roberto Ierusalimschy and Luiz Henrique Figueiredo for our discussions about the past, present, and future of Lua, for their comments on drafts, as well as for their willingness to stand aside and let me present the story as I saw it. I thank the members of the Kepler team and Alta's developers and managers for tolerating a resident ethnographer in their midst.

Analysis of the interviews would be much harder without access to transcription. I thank Siobhan Hayes, Eva do Rego Barros, Rosa Paiva, Eliodora Besser, Patricia Martinez Alzueta, and Mariana Timponi for their work. I thank Marcelo Besser and LoGoS Traduções e Consultoria for organizing the process.

Conversations with Brazilian scholars have helped me better understand the local context and Brazil's history. I am particularly thankful to Paulo Tigre, Ivan da Costa Marques, Sidney Oliveira de Castro, Henrique Cukierman, Antonio Botelho, Nelson Senra, and Simon Schwartzman. I thank the Institute of Economics of Universidade Federal do Rio de Janeiro for hosting me in 2005.

After finishing my dissertation in May 2009, I moved to University of Toronto, where I benefited from support of colleagues and students. I am particularly thankful to Anthony Wensley for his efforts to bring me to Toronto and for allowing me to dedicate time to research. I thank Matt Ratto for helping me orient myself around the Faculty of Information and for his advice on publication strategies. I would also like to thank Brian

Cantwell Smith, Jun Luo, and ginger coons for thoughtful comments on drafts, and Annie Shi for her assistance in copyediting.

The book benefited substantially from the attention of the series editors, Bonnie Nardi, Kirsten Foot, and Victor Kaptelinin. They have supported this work in its transition from a dissertation to a book and suggested many of the needed revisions. I thank the anonymous reviewers of the manuscript who read the book with a great attention and provided valuable comments. I thank Kathleen Caruso at the MIT Press for her thoughtful edits and other MIT Press staff for their work on this book.

Luisa Farah Schwartzman has seen more revisions of this manuscript than anyone else (with the possible exception of Paul Duguid). She has helped me reduce the amount of technical jargon in my writing and increase the number of definite articles. Having a native speaker of Portuguese around and ready to help decipher the most enigmatic passages in my recordings undoubtedly qualifies as an unfair advantage. As two scholars working side-by-side, we have gradually found more and more intersections in our bibliography and have often jokingly argued about who stole whose ideas. I may never be able to fully appreciate Luisa's impact on my work.

I thank my family for their support over the years. My parents have helped lay the foundation for my own globalization projects. My father's global dreams and his ability to imagine his children's future education abroad even before the fall of the Berlin Wall have helped bring me from Vladivostok to California. He introduced me to my first computer, and I often remember our long conversations about the nature of information nearly two decades ago. My mother has contributed greatly to my international adventure by helping me develop a passion for foreign languages.

This book is based on work supported by the National Science Foundation under Grant No. SES-0724707, by the Berkeley Fellowship, and by Yahoo! Research Key Technical Challenges Grant.

A Note on Translation, Quoting, and Pseudonyms

Most of the quotations included in this book are my translations from Portuguese. Readers interested in seeing those quotations in the original Portuguese can find them on the book's web site at http://codingplaces.net/, which also describes my approach to transcription, quoting, and translation. My interviews with the two authors of Lua were conducted in English, while my conversations with "Rodrigo Miranda" often alternated between the two languages. (The companion site identifies the original language of each quotation.)

The book uses two methods to present direct speech to account for the variation in precision with which the speech was captured. I use quotation marks or block quotes for speech that is reproduced verbatim with high confidence. This includes quotations from audio-recorded interviews and electronic communication (email or instant messenger), as well as phrases recorded verbatim in my notes. I use direct speech *without* quotation marks for utterances that actually occurred and closely match what was said, but may not be reproduced verbatim. In some cases, I put such utterances in italic to set them off from the rest of the paragraph.

I sometimes use simple ellipsis ("...") to indicate disfluencies in the original speech, for example, unfinished sentences, short pauses, or breaks in sentence structure. I always use bracketed ellipsis ("[...]") in places where a part of the quotation is omitted. Additional details on the quoting method are available on the companion site.

I use pseudonyms to identify the participants in most cases. (The authors of Lua are the main exception.) I also use pseudonyms for names of several companies and software products. Each pseudonym is shown in quotation marks the first time I use it but appears without quotation marks if it is used again later.

0 The Wrong Place

Why would you come from California to Rio de Janeiro to study software developers? The question was asked in a friendly tone, with just a touch of suspicion. It would not send blood rushing through my veins if not for the place where it was asked. I was stooping in front of a small window, in the midst of explaining to a US consular officer why a Russian citizen born in Vladivostok would be seeking an American visa in Rio de Janeiro, at nearly the exact opposite side of the world from where I was supposed to be applying for it. I was in the wrong place, and a good explanation was due, lest my personal world should suddenly become far from flat. Saying that I had come to Brazil to study *software developers* was a sure way to raise eyebrows further.[1]

I will try to show in this book that we have much to gain from looking at software development in this somewhat unlikely place, and more generally, from looking at high-tech work in "wrong" places. By doing so, we can learn a lot about *place* and its persisting importance in today's "knowledge economy." For over a decade, popular authors have declared that place will soon become unimportant for human activities, as people increasingly gain the ability to communicate and collaborate over distance (Cairncross 1997; Friedman 2006). In the age of the Internet, they have argued, where you are does not matter. Others have countered such claims, pointing out that the world might actually be becoming more "spiky," with a small number of places *growing* in importance as centers of global activities (Florida 2008). Picking the city to live and work in, they say, may be your life's most important decision. If you are in the wrong place, pack your bags quickly and move! And some people do exactly that. For decades, places like Silicon Valley have attracted (and continue to attract) people from all over the world. Eighteen years ago, I myself left a provincial Russian city for Palo Alto. Most people stay close to where they were born, however. This book is about those people, the work they do, and their role in globalization.

My story and analysis challenge both views outlined earlier. I argue that we should neither declare "the death of distance" nor fix our gaze on a handful of "spikes." Instead, we must look at globalization as an active process arising from the combined efforts of many people around the world working daily to defy space, building individual connections to remote places in pursuit of global dreams. To understand globalization we must look closely at such people: at their goals, their struggles, their failures, and their successes. We must pay attention to how their efforts reduce or increase differences between places. And we must look in the wrong places.

Practice and Place

The book looks at people who inhabit simultaneously two different contexts. One of those contexts is defined geographically—a metropolitan area in southeastern Brazil, consisting of the city of Rio de Janeiro that is home to around six million people known as *Cariocas*, and the adjacent municipalities inhabited by an equal number of *Fluminenses*, many of whom commute to Rio de Janeiro for work. The other context is an instance of what I call *worlds of practice*—systems of activities comprised of people, ideas, and material objects, linked simultaneously by shared meanings and joint projects. Such worlds vary in scale, but many of them are global, connecting people and objects spread around the planet. The world of software development is global in this sense, inhabited by around ten million people who are spread far and wide. I argue in this book that global worlds of practice are the key constitutive elements of globalization. In other words, to understand globalization we must look at not just the technologies that enable global communication, nor the structures of global governance. Rather, we must investigate the global "worlds" that form around specific systems of human activity, noting how globalization projects occurring within such systems reinforce each other and produce the overall experience of globalization.

The world of software development makes an interesting context for a study of globalization because it exemplifies its paradoxes like no other field. Software development is often seen as a quintessential example of "knowledge work," a global profession, freed from the constraints of geography by the immaterial nature of its inputs and outputs. Whereas traditional industries convert material inputs into material outputs, and moving those inputs and outputs costs money, "knowledge work" focuses on transforming "knowledge," an entity that can be easily imagined as perfectly mobile—at least as long as our idea of "knowledge" is modeled largely on computer files. And while this crucial resource could in theory be hoarded by a privileged few, in practice it is often seemingly rendered free for all by the collective generosity of "communities of geeks," which Friedman (2006) sees as an example of a broader "uploading" of knowledge.

Given the abundance of uploaded knowledge, engaging in software production seemingly requires little more than a computer, stable electricity, and Internet access—all of which are available in places like Rio de Janeiro even to the relatively poor. Armed with those tools, developers can access vast repositories of code and documentation from across the globe. Brazilian developers sometimes spare no words when describing the significance of the Internet to their work. They speak of it as "the world's greatest library," full of "all the imaginable and unimaginable resources." Developers can use code and documentation found on the Internet to build their own solutions. They can then distribute the products of their labor to people around the world, again using the Internet. Occasionally we read news stories that seem to illustrate the ease of this scenario. For example, in 2009 a seventeen-year-old Moscow high school student built Chatroulette—a video chat system that soon had over a million users from around the globe and was discussed in the news all over the planet.

Such stories, however, must not distract us from another notable feature of the world of software: its stark and persistent centralization. Over the last several decades, the world of software has revolved around a handful of places. One of those places—Silicon Valley—has in fact become a textbook example for illustrating the idea of regional clustering of industry. In addition to being home to a large number of software practitioners, Silicon Valley and the greater San Francisco Bay Area also serve as a base for some of the world's largest and most successful IT companies that control the work of developers around the world. Together, market capitalization of IT companies headquartered around San Francisco comprises over a third of the world's total (see chapter 4).

This concentration of valuation is indicative of the difference in the *kind* of software work that gets done in different places and the geography of control over software work. Many of the developers working in San Francisco and in some of the other centers of the software world apply their efforts to software intended for broad use, which would, if successful, bring their companies big rewards. Such rewards can be both financial and symbolic: the successes of Oracles, Apples, and Googles make up a good part of the global software lore. Software developers working outside such major clusters recognize the preeminence of remote centers. Stories such as those of Chatroulette often have a little-noted ending: the developers moving to San Francisco Bay Area or selling their venture to a company based there.

The practice of software development thus appears to be simultaneously remarkably placeless and starkly placed. This paradox can perhaps be grasped most clearly by considering the case of Google, the company whose search engine is often mentioned as the greatest "leveler" by Brazilian programmers, but which itself arose—and most likely could only have arisen—in a highly predictable place, biking distance to Silicon Valley's Sand Hill Road.

By most counts, Rio de Janeiro is a peripheral place in the world of software. In terms of sheer numbers, Rio de Janeiro likely has about one-tenth the number of programmers of the San Francisco Bay Area; in terms of IT valuation, the difference between the two regions likely approaches a factor of *one thousand* (see chapter 4). Developers who work in Rio usually dedicate their efforts to the smaller problems faced by local organizations. The most successful address the needs of Brazil's national market (though many usually find that such work is better done elsewhere, in the larger São Paulo). "This is not Silicon Valley," Rio developers often explain when talking about the possibility of taking on more ambitious projects.

Yet it is precisely this peripheral position in the remarkably centralized world of software development that makes Rio an interesting place for looking at knowledge work. After all, while the software developers working in Rio are fewer than those in Silicon Valley, the overwhelming majority of people who write software do so in places that are more similar to Rio than to Silicon Valley.[2] To understand the truly exceptional position of centers such as Silicon Valley, perhaps it helps to spend some time contemplating the periphery. What do software developers *do* in such places? Why do they do it? Answering such questions will help us better understand the nature of ties that bind together the world of software and today's global society.

To make sense of the paradox between software's seeming independence from geography and the centralization of its production, we could try to understand why software development remains so concentrated in the era of unrestricted knowledge flows—a popular road that seems to almost inevitably lead one to ask what is *wrong* with all the places that fail to produce a thriving software industry. I touch upon this question at several points in this book. For most of it, however, I take a different approach. Instead of assuming that technical knowledge is naturally fluid and trying to understand what barriers keep software development so concentrated, I take the concentration as a given and seek to understand how the practice of software development moves in space *at all*, investigating the work that is needed to establish this practice in new places. How the seeming universality is achieved *in spite* of this geographic concentration then becomes one of the key questions.

In doing so, I put aside the term "knowledge" for the sake of another one: "practice." To understand how knowledge comes to new places, we must look at it in conjunction with all other things that must be in place to support its power—the social arrangements that provide the "tracks" along which technical knowledge can travel (Latour 1987). While this expansion of scope could be done by arguing for a broader notion of "knowledge," I switch to a different term partly to draw on the rich body of social theory from which I borrow the concept of "practice," and in part because I feel that the tendency to think of "knowledge" as something akin to the content of computer files is so strong today that I cannot expect the reader to ever fully leave behind this unfortunate metaphor.

The concept of "practice" provides us with a useful analytic layer between the more abstract, propositional notions of knowledge and the messy details of daily life. As I explain in more detail in the next chapter, I understand "practice" as a system of activities, a collective way of doing certain things, or a system of "doings and sayings" (Schatzki 1996). A practice maintains continuity through a mutually sustaining relationship between patterns of interactions, material resources, and shared systems of meaning. Looking at the practice of software development, I thus look at the *doing* of software development, the people and groups that engage in this doing, and the relationships between them. I also look at how such doing interacts on the one hand with ideas and discourse, and on the other hand with the material elements of the practice. This nexus of relations creates a *context* for individual actions, a context that individuals can "inhabit" in ways that can be likened to how they inhabit physical places, and to which they can have commitments—commitments that must be balanced with those to the local place and the national community. Such contexts are bounded and often named. The developers sometimes talk about being in "the world of software." For this reason, I describe such systems of activities as *worlds* of practice.[3]

Focusing on activities, and especially on *systems* of activities, makes it easy to see why the practice of software development would cluster in a handful of places, since it helps us recognize the many different pieces that would need to be put together to re-create the practice in a new place. For someone who adopts this perspective, the problem becomes that of comprehending how a living practice could *ever* move to new places. To put the same question differently, we can ask how "uploaded" knowledge and other elements of the practice, removed from their original context, are put together and made to work in a new place.

My discussion of practice in place focuses on several themes. The first is *the process of disembedding* and *reembedding* (Giddens 1991) involved in its reproduction across space: people engaged in a practice that is based some-where else often have to reassemble the practice around imported elements, substituting for missing pieces what happens to be available. (And if they want to get involved more centrally, i.e., *extending* the practice, they will have to find ways to thoroughly disembed their own innovations, to make them mobile and useful in the remote places where the practice is stron-gest.) The second theme is *the cumulative and parallel nature of the reproduc-tion process*. I look at the local practice of Brazilian software developers as a partial reproduction of the American software practice and frame my obser-vations as a particular moment in the history of this practice—a moment when many elements have already been brought in and reassembled (hence the need to look at history in chapter 4), while others are still missing. I also look at this reproduction as one of many parallel efforts to re-create foreign practices. Third is the theme of *a "diasporic" situation of the peripheral prac-titioners*, who engage simultaneously in two cultures: the local mainstream culture and the globalizing world of the practice. (Those engaged with the practice at its centers may face this issue as well, but the gap between the two worlds is usually not as wide.) In particular, I look at how commitments to those two cultures come in conflict and how such conflicts are nego-tiated. Closely related to this is *the complex relation between individual and collective efforts of reproducing foreign practice*: local practitioners must often decide whether to cast their lot with their local colleagues or focus on their individual connections to remote centers. The fourth theme is *the interaction between the cultural and economic layers of the practice*, and the need to look at the two simultaneously, considering the situations when one of those layers is present and the other is missing. Finally, I stress the importance of paying attention to *actors' reflexive understanding of the world*, the possible futures they can imagine individually and collectively, and the factors that influence this imagination (Giddens 1979; Appadurai 1996). Together those themes provide us with a view of globalization that highlights individual agency of peripheral actors, situating their actions in the context of cultural and economic structures, while also showing how their individual attempts to engage in global systems of activities add up, collectively and over time, to create the seeming universality of global practice.

By bringing to light the work that peripheral practitioners must do to give software development its seeming universality, I hope to offer them the credit they deserve (and all too often deny themselves), touching upon the question of why software development remains centralized. While I do not see this centralization as a puzzle per se, I do believe that there are many explanations that are wrong and self-serving, and that such explanations may themselves contribute to the persistence of centralization. The discussions of the geography of software work (or other types of "knowledge work") and the feasibility of developing "the next Silicon Valley" in this or that place quite often arrive at the importance of attracting "smart people" (e.g., Graham 2006). While smart people are undoubtedly important for a successful software industry (as for many other types of work), researchers and policy makers sometimes seem too quick to assume that places that lack strong software industry lack smart people. In fact, if one assumes that technological knowledge flows naturally between capable minds and is sufficient for the re-creation of a knowledge industry, then the concentration of software development in a handful of places would seem to imply that other places lack smart people, smart governments, smart investors, or all of the above. Unfortunately, such judgments are often internalized by the peripheral actors themselves, who might sometimes consider themselves an exception to the rule, but too often assume that the mediocrity of their fellow citizens limits what they can achieve. Highlighting the work that went into bringing about the current state of affairs, and the achievement inherent in that, I hope will present a brighter picture and in turn facilitate local cohesion.

I also intend to show how such peripheral work contributes to the continued dominance of remote centers. Like many other knowledge products, software production is characterized by strong network effects: software that is used becomes more useful and will often gain in popularity because of its popularity. This is often especially true for open source software (which I discuss in the next section), where products that are widely used often actually become *better* as they attract more contributions. By fixing their gaze solidly on foreign technology and investing efforts into making it work locally, peripheral developers often deny to local projects the attention that such projects may need. Such lack of attention and, more important, lack of *trust* in local projects is ironically the opposite of what has been credited for making Silicon Valley the success that it is—the strong networks of personal relations and personal trust (e.g., Saxenian 1996). Unlike in the San Francisco Bay Area, in Rio being local carries a stigma and the local place works against the practitioners. The local developers are thus themselves involved in replicating the asymmetries from which they suffer.

Such observations should not be interpreted as suggesting that peripheral participants and regulators should either turn away from foreign technology or desist altogether in light of the challenges. As Brazil has learned in the past, isolationism can be a dangerous strategy and nuanced solutions are needed. I do not make specific policy recommendations, but I invite policy makers to follow me on a visit to a world that they govern (in part) but do not always understand, to see the challenges faced by people who inhabit this world and to consider how helping them face those challenges may contribute to the larger developmental agenda. I hope in particular that the case of Kepler, an open source software project described in chapter 8, read together with the two alternatives to the approach Kepler exemplifies (chapters 5–7), will be useful for thinking about innovation policy.

Peculiarities

Understanding the way a universal practice is made to work in a concrete place requires looking at the many peculiarities of that place: the specific configurations of resources that are available to the actors who inhabit that place and the specific history that has led to those configurations. It is for this reason that I focus on a single city and present it as something concrete, rather than sampling software developers from a wide range of places and losing the concreteness. While looking at one specific place, however, I seek to show relations I believe exemplify the patterns we can find in many other places. Every place has a history and every place has a local context. In every place concrete work must be done to turn abstract knowledge into a living practice.

While I believe the patterns I explore in this book could have been shown using many other cities, the choice of the specific place can make a difference. Different degrees of peripherality would bring into focus different parts of the reproduction process. Focusing on a place where the practice of software development has yet to take root would help us see the earliest steps in this process, but would shorten the history available to exploration, leaving us to imagine all sorts of possible scenarios for the future. Picking a place that is secondary today but could have become the main center of information technology had the history of the twentieth century gone just a little differently (e.g., Cambridge or Berlin) would highlight the importance of contingencies, but would give us little insight into the future possibilities. I believe that my choice of place gives us a good balance: a city present on the world map, yet not quite one of the "global cities"; in a developing country that seems to be gaining momentum, yet doing so at a pace that allows for some reflection; and with a history of IT policy that goes back a few decades—putting some of the most important events in this history far enough back to allow for critical analysis.

In addition to the peculiarities of Rio de Janeiro, two other aspects of the book may strike the readers as unusual and thus call for a brief introduction.

Free / Open Source Software

The book focuses disproportionately on a specific form of software practice known as "open source" or "free" software development. Although those two terms vary substantially in connotation, both refer to software that is distributed in a manner that allows the recipient to modify it, and then redistribute it to others without paying royalties to the original author. While such distribution of software has been common since the earliest days of software, it has come to particular prominence since the development of Linux, an open source operating system, in the 1990s.[4] In recent years, the development of free / open source software has attracted substantial attention from social scientists, including sociologists and anthropologists who have often looked at it as a political and cultural movement (e.g., Kelty 2008) and economists who have looked at efficiency gains associated with this form of software production. In this book, I look at cases of open source software development through the lens of *practice*, highlighting the interrelations between culture and material production, and positioning open source within the context of the global world of software practice.

Open source software development presents in perhaps the clearest form the paradox between placelessness and centralization described earlier. Open source communities are intentionally open, and the apparent generosity of those "communities of geeks" provides much of the motivation for Friedman's discussion of "uploading" as one of the key factors contributing to the "flattening" of the world. Such communities are also remarkably dispersed and rely predominantly on computer-mediated interaction, with members often having little idea where on the planet other participants happen to be. At the same time, however, the geographic concentration of those communities rivals that of the software industry, with rare projects that originate in "wrong places" often quickly moving their centers to the West Coast of the United States. The global culture of such communities is based largely on the "hacking" culture that originally developed in American universities.[5] Their practices are today supported by business models pioneered by American companies and optimized for the situations they face. English is almost always the working language of such communities, even as they might strive to create software products that support every last script on the planet. As I will try to show, participation in open source projects involves a complex negotiation of culture, language, and geography, and is often *harder* than engaging in other forms of software practice, since it requires *more* fluency in foreign culture and demands *more* of the resources that may be hard to find in places like Rio de Janeiro.

Open source development contributes to globalizing the practice of software development. It is important, however, to avoid trivializing this relationship and to consider the local work that mediates it. Open source development creates a new opportunity—and a challenge—to participate in projects based far away. To take this opportunity and respond to this challenge, however, Rio developers must learn quite a bit more about foreign practices and find more of the missing pieces of the practice.

On a more abstract level, open source development also simply represents a *new* way of developing software, and thus highlights the challenge of keeping up with the evolving practice based far away—what we could call "synchronization work." Looking at how Rio developers respond to this challenge may therefore help us understand how people engaged in other worlds of practice respond to changes that take place in those worlds.

Lua

Several chapters of the book look closely at a particular open source project that would be unusual by most measures: Lua, a programming language developed in Rio de Janeiro that has recently gained substantial global popularity around the world—in particular, among software companies based in California. For example, Lua was used extensively in *World of Warcraft*, a networked computer game played by over ten million people (a number sufficiently high to secure an entry in the Guinness World Records), and more recently in *Angry Birds*, a game that was downloaded over one hundred million times in its first fifteen months. Lua has also been used in products made by Google, Adobe, Microsoft, Verizon, Cisco, and other technology companies.

Lua's global success is surprising, not the least to those people in Rio de Janeiro who are familiar with the scale of its use abroad. It is particularly stunning when we consider the powerful network effects that ensure that the number of programming languages in common use remains quite small. Lua is the only entrant into this exclusive club from a developing country.

Lua's position in Brazil, however, presents us with an even larger puzzle. Almost no local companies make use of Lua in their products. Lua's large and active community interacts primarily in English. Software developers in Rio de Janeiro who wish to learn Lua can do so using a book written by one of the authors of the language (a professor at a local university), but they will need to read the book in English, because no Portuguese translation of the book is available. Unless they know the author person-ally, they will likely also need to order the book from Amazon.com and have it shipped from the United States, since no Brazilian bookstores carry it on their shelves. Lua's global success has so far done little to rescue Rio de Janeiro from its position as a "wrong place" for developing software. Programming in Lua has just become another activity that is better done in Silicon Valley.

I present Lua as a case of a particular strategy of engagement with global technology: a focus on global connections, in the name of which local link-ages may have to be sacrificed. I show the reasons for such disengagement from the local context, as well as some of the efforts to reconnect Lua to Rio. I discuss the strengths and weaknesses of this approach, presenting a range of perspectives on Lua's past, present, and future. I contrast this case with two others: the localization of global technology by a successful IT firm in Rio (chapter 5) and an attempt to bridge the gap undertaken by a government-funded open source project aiming to make use of Lua locally (chapter 8).

The Project

This book is based on an ethnographic project—an attempt to understand the experience of a group of people through an extended engagement with them. In my case, this meant a combination of over one hundred inter-views, extended presence in places where software work was being done, and at times active engagement in the members' projects.[6] As Van Maanen (1988) points out, ethnographers use different approaches to present their observations. Some tell "realist tales": accounts that simply present what happened, taking as given the ethnographer's ability to know and to inter-pret it. Others tell "confessional tales": accounts that focus on the observer as much if not more than they do on the observed. They normally do so out of the realization that the observer inevitably influences what is observed, and that the process of observation and interpretation is often fragile and its success is contingent on many factors. The inclusion of the observer in the account helps the readers better understand what was observed by being told who did the observing and how. It also helps the ethnographers consider their own biases, as it encourages them to think more closely (and explain to their readers) about their own role in the events.[7] (It also, as Van Maanen points out, helps establish the ethnographers' authority by show-ing that they have gone to places where the readers have not been.) Though I include myself in the account whenever appropriate, I avoid the extremes of confessional ethnography, finding it potentially distracting from the larger points that I want to make. In particular, the order of the chapters reflects the theoretical logic of the book rather than the chronology of my fieldwork. To compensate, I present a brief "confession" in this section.

In the summer of 2003, after spending three years working as a software developer in Mountain View, California, the heartland of the region known worldwide as "Silicon Valley" (but referred to locally as just "South Bay" or "the peninsula") and before starting my PhD program at Berkeley, forty minutes away by car, I spent a month in my hometown in Vladivostok, Russia, on the other side of the Pacific. (As I learned a few years later, this city is known to many Brazilians primarily as a base for attacking Alaska in *War*, a board game based on the American *Risk*.) While there, reconnecting with old friends and meeting new people, I saw a world that I had started to forget during my years in California. I was in a place that seemed in some ways quite provincial, yet at the same time was much more global than Mountain View. One could not find in Vladivostok California's diversity of cuisine or languages, yet the existence of the external world was much more apparent than it ever was in California. Many of my conversations revolved around places outside Russia—in particular, the United States, which seemed to be visible from Vladivostok in the way no country is from California. I started developing an interest in understanding how people who work in "peripheral" places maintain ties to the places they consider more "central" to their field. Trips to Brazil and Finland the following year solidified this interest.

By September 2004 I had decided to focus my dissertation research on software developers in Brazil and their access to software knowledge from the foreign centers of software practice. At the time, it did not occur to me to ask whether Brazilian software developers were in fact in a place where locally generated software knowledge was in short supply and whether they actually tried to access knowledge from places such as Silicon Valley. Both assumptions turned out to be correct—the developers I later interviewed typically saw Rio as no match for Silicon Valley as far as software goes, and they most certainly did seem focused on keeping up-to-date with what was happening abroad. Such assumptions, however, hid many of the questions that later came to dominate my thinking and to which I will turn shortly.

After another year at Berkeley, spent learning Portuguese and reading social theory and economics, I arrived in Rio de Janeiro in June 2005 for a six-month stay and started building my sample of "software professionals." I defined the term loosely, including in it people who were trained to write software, regardless of whether they actually wrote it as a part of their job, and people who actually wrote software, regardless of whether they were trained to do so. (I later switched to the term "software developers" to avoid the presupposition that software developers are "professionals.")

My sample combined elements of a "theoretical sample" and a "snowball sample." The term "theoretical sample" describes an approach to sampling that involves the researcher seeking "cases" they hope will challenge their preliminary assumptions and lead to further development of the theory (Glaser and Strauss [1967] 1999). Such a sampling technique often aims to increase the diversity of the sample, in order to compensate for its small size. A common way of building such a sample is by asking interviewees to recommend additional people who could be interviewed (a "snowball" technique), either specifically asking for people matching certain characteristics or selecting them from among the nominees. In my case, I attempted to include among my interviewees every type of software developer I could identify, "oversampling" atypical individuals. For that reason, I made sure to interview not only developers graduating from top universities, but also their professors, continuing my quest for the ultimate "alpha-geeks" until I interviewed two of the authors of the Lua programming language, to which I dedicate chapters 6 and 7. I similarly attempted to include developers with as little education as I could find. I interviewed people from a range of work environments: small companies, large local companies, multinationals, university research labs, people officially employed by their companies and those hired as contractors, employees of the public and private sector. I talked to people of different ages, and I made an attempt to include women in what otherwise was turning out to be a heavily male-biased sample.

I came to Rio with basic knowledge of Portuguese, though not quite ready to conduct interviews in Portuguese comfortably. My earliest interviews were thus conducted in English, while I was also taking private classes to better prepare for interviews in Portuguese. I started conducting such interviews in the beginning of my second month, at first resorting to Portuguese only when talking to interviewees who could not speak English. As my Portuguese fluency improved, I conducted more interviews in Portuguese, eventually using English only with the developers who spoke fluent English and preferred to talk to me in it. This awkward start and the subsequent change in the language of the interviews turned out to be a blessing in disguise. My own struggles with Portuguese made me somewhat more sensitive to my interviewee's struggles with English (and more appreciative of their successes with it), while alternating between English and Portuguese exposed me to the different discourses invited by each language.

During my second month in Rio, I learned a methodological lesson that greatly affected the rest of my project. Most developers that I had interviewed up to that point assured me that they never discussed technology outside work, which I found quite surprising. I brought this up during an informal post-interview chat with a developer who did mention discussing technology and work with friends. He suggested that the other interviewees were simply not willing to admit it. Talking about work, he explained, was simply not considered cool in Rio—young men are expected to talk about soccer and women, not computers. He assured me that my other interviewees did talk about technology with friends, and that I just had to know how to ask. As I soon came to realize, small differences in wording and intonations did indeed affect greatly the interviewees' readiness to talk about talking about technology. The incident also made me realize, however, for the first time, the subtle incongruence between the local culture and the seemingly global software practice. Furthermore, it led me to start paying attention to not only what my interviewees were telling me, but also why they were telling me that, as well as to things that were unsaid or sometimes half-said. (Another point that I learned in this and other similar interactions was the importance of drawing on "ethnomethods"—developers I interviewed in Brazil became a great source of explicit advice on how to interview other Brazilians.[8])

In late August 2005, when discussing my plans with a senior official of the Ministry of Science and Technology, I got reprimanded for trying to understand Brazilian reality in isolation from Brazil's history. I took this criticism seriously. Though my investigation into Brazil's history and its relation to the current practices did not come together until after my return, I did use my visit to discuss my interests with a number of scholars affiliated with the Federal University of Rio de Janeiro, including some who observed firsthand Brazil's technology policy in the 1970s and 1980s—or even helped shape it. Those conversations helped me gain a better understanding of Brazilian history from a perspective that is not very popular today, especially among the younger of my interviewees. This perspective gave me a point of comparison that helped me question the idea of "global technology" and start looking at how the global nature of technology is constructed through local work.[9]

In January 2006 I returned to Berkeley to analyze my data and to prepare for my qualifying exam before another five-month trip to Rio. During that time my interest increasingly shifted from the mechanics of how my interviewees kept in touch with foreign technology to the tensions and contradictions in some of their accounts. I came to see those contradictions as reflecting the underlying conflicts between their commitments to the local place and to the "global" (but often also quite foreign) technological practice. I also started recognizing in those tensions the different images of the world that the developers had.

In September 2006 I exchanged a few email messages with "Rodrigo Miranda," one of my first interviewees in 2005, a coordinator of an open source project called "Kepler." The project aimed to build a web development platform based on Lua, the programming language developed in Rio de Janeiro that I referred to earlier.[10] When I mentioned to Rodrigo that I was planning to return to Rio in early 2007, he asked me if I would like "to participate in the Lua adventures," adding that he might be able to find funding to pay me to work on some parts of Kepler. I declined the job offer but took time to learn more about Kepler and Lua—projects that I earlier treated as too atypical for serious investigation. As I learned more about them, I found myself puzzled and surprised at every step. I was also starting to get a new understanding of the more typical cases. I then decided to dedicate half of the second phase of my fieldwork to Lua and Kepler, reserving the other half for a study of a more typical case—some company building custom web applications for local clients, using Java, a popular programming language and a software development platform by Sun Microsystems, a California company.

In February I joined the Lua mailing list, spent some time reading its archives and did six interviews with Lua users in California. I then went to Rio to start a new round of fieldwork, having already secured not only Rodrigo's invitation to study Kepler and the Lua team's blessing for studying Lua, but also a desk in Rodrigo's office at "Nas Nuvens," a company that sponsored Kepler. I thus jumped into my study of Kepler right away, leaving my study of a "typical" company for the later part of my stay. As it turned out, I arrived at the right time: Rodrigo was about to try a new approach to the project that would aim to "open" it in order to draw in a larger number of remote participants.

Despite having seemingly open access to the project and getting a chance to meet most of the participants early on, I soon confirmed my suspicions that mere physical observation does not go very far when studying software work: one mostly gets to see people staring at their screens, typing, and occasionally swearing. Such observation gets even more complicated when the participants do their work in different parts of the city, which cuts the amount of time dedicated to water cooler conversations even further (replacing them, e.g., with instant messaging, a more private medium). Without literally looking at the developers' monitors over their shoulders, both at work and at home, and keeping track of their solitary work, private emails, and instant messenger conversations, cell phone calls, and face-to-face chats, one can hardly see all the work that goes into the creation of the software project.[11]

I tried to compensate for this with interviews, but my conversations with the developers often seemed too removed from what they were actually doing. In fact, after a few weeks, I began to doubt whether anything was actually even happening. I decided to start helping Rodrigo with the project's web site, but felt that even this was giving me too secondhand of a view. Our discussions of the web site, however, soon arrived at the conclusion that we wanted to run it as a wiki—a web site that allows visitors to make changes to the content. After we went through a number of options for wiki software, I made a fateful decision to write my own wiki in Kepler, which was after all a platform for developing web applications such as wikis. Even though writing a simple wiki only took a few days, it immediately changed my place in the project. As the first public application built on the platform, the wiki generated immediate interest—and immediate demand for improvements. As I started spending time making changes, my conversations with the developers changed. I was now one of them, a member of the project and the larger "Lua community."

I found in such active participation an answer to many of the problems of studying software work that troubled me at first. While no method can reconstruct the project in its entirety, active *participant* observation provided me with a partial solution: a situated and integrated picture that weaved together *some* private emails and instant messenger conversations, *some* late night conversations over pizza, and quite a few hours alone in front of the monitor making sense of debug traces.[12]

Such engagement also created a number of challenges. While my own technical background proved a blessing because it allowed me to get engaged, I soon came to face the challenge of getting involved without "going native." A certain degree of resocialization is of course a crucial aspect of the ethnographic experience; hence, many ethnographers believe in doing ethnography far enough from home to achieve isolation from the home environment (see Van Maanen 1988). Too much involvement, however, can limit time available for reflection. It also raises serious questions of commitment. I got asked, on quite a few occasions, whether my participation in the project was "serious" or "just a research project." To be a participant was to be involved in a "serious" manner, treating the activity as meaningful and important on the same terms as the other members. Faced with this choice, I decided to get involved seriously. This led to a struggle to maintain balance between my life as an ethnographer and my life as a Kepler developer, but in the end I felt it was worth it.

In traditional ethnography, the obvious need to physically return home provides ethnographers with a natural end to the involvement—and hopefully keeps them from making unrealistic commitments before that. Virtual projects done over the Internet create an opportunity—and in the view of some members an *obligation*—for the ethnographer to maintain commitment to the project through continued remote participation. Since leaving Brazil in August 2007, I have stayed involved with the project, following it through its ups and downs, finding it impossible to disengage from it even after Rodrigo himself decided to move onto other things.

In June 2007, I moved my base from Rodrigo's office to the office of "Alta," a software company building Java web applications for local clients. My time at Alta was shorter than my engagement with Kepler and also substantially less participatory, because I continued to be involved with Kepler and conduct interviews related to both Kepler and Lua. As it soon became clear, my commitment to Alta was insufficient for the company to depend on me—especially when its obligations to clients were at stake. To say it differently, my participation would not have been sufficiently "serious." I had to settle for a relatively passive role: spending time in the office, chatting with the employees, conducting sit-down interviews with them, sometimes watching their work over their shoulder, poking around in the code repository, but not actually doing their work together with them. Despite such limited participation, the six weeks spent in Alta's shiny office provided me with an opportunity to better understand a different, and in many ways a more typical, work environment.

I returned to the United States in August 2007, bringing with me 150,000 words of field notes, not counting notes and recordings for over a hundred interviews.[13] Over the course of the following years, I sifted through this material and theoretical literature, looking for ways to put the two in a conversation with each other. The next section outlines the result of this process.

The Chapters that Follow

This introduction is followed by eight chapters and a conclusion. Chapter 1 lays a theoretical foundation for the book, outlining with more precision analytical concepts employed later in the book. (Readers who would

prefer to avoid a strong dose of social theory upfront may consider starting with chapter 2, perhaps returning to chapter 1 later.) I argue that to
understand Rio developers' engagement with "software development,"
we must conceptualize software development as "a world of practice." I
show how Giddens's theory of structuration can be extended to analyze
the spatial expansion of worlds of practice and how such analysis can be
applied specifically to software development. In chapter 2, I add some ethnographic flesh to the theoretical skeleton developed in chapter 1, taking
a slice through many of the contexts explored in the book while focusing
on a particular theme: the use of English and Portuguese by Rio software
developers, which illustrates the developers' position between the world of
software and the local world of Rio de Janeiro.

Chapter 3 explores developers' early steps toward the software profession, looking at biographies of a small number of developers. I first show the
adolescents' cultural entry into the world of computer "nerds" that often
precedes the engagement with labor markets. I then turn to the transition
from the cultural to the economic engagement with the world of software
practice. Throughout this chapter, I look at how neophyte software practitioners build both local and global connections, entering the world of
software simultaneously *from* and *in* a peripheral place. Chapter 4 switches
from the situated perspectives of Rio's young nerds to a broader look at the
world of software development as a whole, outlining its history and looking at its current geographic organization, pointing out a strong asymmetry
between the kind of work that is done at the "central" and that which is
done at the "peripheral" sites. I then turn to the history of computing in
Brazil to explain the local politico-economic structures encountered by the
young nerds seeking to convert their passion for global software into a local
career. Despite Rio's relative isolation from the larger centers of software
production, a myriad of ties link the city to other parts of that world. This
chapter shows how some of those ties were built historically and the work
that went into their construction.

Chapter 5 starts a discussion of the opportunities faced by Rio software developers today by looking at Alta—in some ways a typical software
company in Rio de Janeiro, engaged in building local applications using
software produced in California. Chapter 6 turns to a very different kind
of project: Lua, the globally successful programming language developed
in Rio de Janeiro that I introduced earlier. I first discuss Lua's use abroad,
drawing on my interviews with users of Lua in California. I then present the
history of Lua, focusing on the ways in which Lua had to gradually separate
itself from the local context to achieve its global success. Chapter 7 turns

to Lua's relationship to Rio de Janeiro and Brazil in the recent years. I show the continued tensions between Lua's adoption in Brazil and its status as a global programming language. I present a number of conflicting—and often *conflicted*—opinions on the possibility of viewing Lua as a "patriotic" artifact and a potential vehicle for local development. Chapter 8 looks at Kepler, a project that aims to bridge the two different worlds: the mostly local world occupied by Alta and the mostly global community inhabited by Lua. The themes of the book are summed up in chapter 9.

1 Global Worlds of Practice

This book aims to understand the nature of globalization, and in particular the nature of the globalization of software work. In my approach to globalization, I start with two premises. The first premise is that globalization is a real phenomenon, and quite likely one of the most important dimensions of the set of transformations taking place in today's world. The second is that we cannot understand globalization just as a matter of space ceasing to matter. Contrary to pundits' pronouncements over the last decade (Cairncross 1997; Friedman 2006), distance is not dead and the world is not "flat." In fact, as many authors have argued, place might be becoming more important than ever before. (See, e.g., Sassen [1994] 2006; Florida 2008.) Understanding globalization therefore requires close attention to *local place*. Yet, we cannot understand it by looking at individual places in isolation. Globalization means a growing importance of *global contexts* that cut across local places. The system of relationships that comprise global software development represents one such context. To make sense of globalization, we must look at such global contexts *in their relationships to local places*. We must note, in particular, how global contexts get connected to each place and how they penetrate (or are drawn into) local processes. For understanding the globalization of technical work and knowledge, a particular kind of global context is crucial: I call it *global worlds of practice*. This chapter develops this notion as a theoretical counterweight to the idea of place.

I use the term *worlds of practice* to refer to systems of activities comprised of people, ideas, and material objects, linked (and defined) simultaneously by shared meanings and joint actions. Each of such systems represents, to quote Schatzki (1996), a "temporally unfolding and spatially dispersed nexus of doings and sayings" (89). In other words, a world of practice involves a system comprised of material actions ("doings"), as well as meanings and signification ("sayings"), that maintains its regularity across time and space. Of course, as we talk of "doings and sayings," we need to remember that the "doings" connect people and material objects, while the "sayings" connect people, objects, and often documents. In other words, the doings and sayings are not the sole elements of a world of practice, but rather the relations that define it.

How such systems of doings and sayings are reproduced in time is an instance of a larger question that has occupied social theory from its early days: the problem of social order—the question of why social systems maintain continuity over time. I believe this question is best answered by drawing on a body of sociological thought known as "theories of practice," as I explain later in this chapter. How such systems are replicated and synchronized across *space* is a problem that has attracted less attention, though it relates closely to a cluster of theoretical challenges that is emerging as central to twenty-first-century social science and concerns the nature and mechanisms of globalization. After introducing the concept of worlds of practice, I focus the later parts of this chapter (and the rest of the book) on this second question, the problem of space.

Worlds of practice vary in the scale of their spatial dispersion. Some of them can be confined to specific places. Many of them, however, are global, connecting places spread far and wide. Being global, however, does not mean being omnipresent or homogeneous. It also does not mean being disconnected from local places. Rather, it means being connected in concrete (and often very different) ways to many places dispersed around the world. Each place must be connected one by one, often through a long process that requires much work on the ground. To borrow an analogy from Latour (1987), a global world of practice can be likened to a railroad network. Once the tracks reach a particular place, people who reside in this place may become members of the radically different context created by the railroad network. Some will gain access to new resources that they can "mail-order" from far away. Some will have new resources used against them. But before all of this can happen, the tracks must first be laid. And they cannot be laid without substantial *local* work.

Establishing a connection between a local place and a world of practice is not a trivial process, and the railroad metaphor does not quite do justice to its complexity. A practice is a system of many interconnected elements; it cannot be simply copied-and-pasted from one place to another. Instead, this process is often best understood using Giddens's (1991) concepts of disembedding and reembedding. First, some elements of practice—people, ideas, tools—must be dislodged from their original context, changed so as to become mobile. Such mobile elements then arrive in a new place, but do so as isolated pieces, disconnected from other elements that gave them power within the original system. To regain this power, they must be reembedded—become a part of a local system of doings and sayings. This usually means that elements brought from afar would need to be made to work with those of local origin, many of them repurposed or pulled out of extant systems. The resulting system will be an assemblage of ill-fitting parts "hacked together," to borrow a programmers' term. In chapter 4, for example, I look at how computing technology was first brought to Brazil in the 1960s, and some of the challenges and "hacks" involved in fitting US-built devices into the local context of Rio de Janeiro.

The work of disembedding and reembedding does not merely *happen*—it is *done* by specific living people. We must therefore understand why and how they do this work. I argue that to understand the process of reproduction and later synchronization of practice, we need to consider that worlds of practice have two sides: a cultural and a material. We must consider simultaneously how participants' actions are guided by systems of meanings as well as by existing configurations of material resources and power. The mutually constitutive relationship between those two sides of practice becomes particularly important when we consider the reproduction of practice across space, examining how imported ideas shape local resources and how imported artifacts are used to shape ideas. In the case of software, we must take seriously the practitioners' claims to be "in love" with their line of work while also remaining mindful of how software development fits in the global economic system.

The local bricolage of foreign and native pieces that results from this process of reproduction may sometimes lead to a new practice that would split off from the original. In many cases, however, the local practitioners will choose to cultivate their ties to the remote places from which the practice was brought, looking to such places as both sources of additional elements and sources of legitimation. They may often find that their continued ability to engage in the practice *locally* and their access to *local* resources requires being recognized as legitimate representatives of the *global* practice. To quote an artist interviewed by Levine (1972), "If you want to show [your art] in Chicago, you must move to New York" (298). Those unwilling to move must put additional efforts into securing their global credentials by other means. Their daily work toward this end will help ensure the growing similarity of local practice to the remote originals. The increased similarity of context in turn will facilitate migration of elements.

Even when the tracks are in place, however, not every station along the railroad is the same. Some are hubs; others see one train per week.

Globalization *links* places, but it does not *equalize* them. Often, it highlights the differences. A small town without a railroad may be a center of its own small world. It must be connected to a center to truly become the periphery. Being peripheral often means being connected tentatively and being dependent on resources that can be withdrawn. And while in some cases, the peripheral participants may hope to become the new center of the practice, quite often peripherality remains an important structural constraint on the development of the practice in a peripheral place—a constraint that is in part perpetuated through the efforts of the local practitioners themselves.

In this chapter, I develop the concept of "worlds of practice" by starting from the notion of "communities of practice" that has become a popular way of thinking about reproduction of knowledge. I point to a number of problems with thinking about software developers as "a community," arguing that the conceptual strength of the idea of "communities of practice" lies in the notion of "practice." I next attempt to unpack some of the conceptual apparatus hidden behind this term, drawing primarily on Giddens's (1979, 1984, 1991) theory of structuration. I then discuss how the theory of structuration can be applied in particular to the problem of reproduction of practice across space, and the importance of paying attention to the global imagination of local actors. I also return to the notion of "community," arguing that, while the idea of community should not be overemphasized, worlds of practice are *bounded* and their members do possess a certain collective identity. Their boundaries and the boundary-making processes are important for understanding the process of reproduction of practice, because practices are not reproduced in isolation, but as a system of practices. This helps explain how a practice can be established in a new place. Finally, I discuss how peripherality affects local practice once the practice is established.

Communities, Networks, and Worlds of Practice

In the late 1980s, a group of researchers from Xerox PARC, a Palo Alto research laboratory, established the Institute for Research on Learning (IRL), a new research center aiming to rethink the fundamental concepts of education. Considering PARC's fame for pioneering in the 1970s much of the technology that defines today's computing experience, one could have expected IRL to make an impact on the field of education through an array of educational software and gadgets. Instead, this impact came in the form of a book that questioned the notion that learning can be understood as acquisition of knowledge. *Situated Learning: Legitimate Peripheral Partici-pation* (Lave and Wenger 1991) argued that learning must be seen not as a matter of transfer of knowledge from the instructor to the learner, nor even as the learner's "construction" of knowledge, but rather as a matter of the learner's deepening engagement with "a community of practice," which the book described as "a set of relations among persons, activity, and world" (98). In other words, learning cannot be understood as anything other than the process that leads a novice to become a successful partici-pant in some collective activity.

In the years after the publication of *Situated Learning*, the notion of "communities of practice" gained substantial currency, especially in the organizational studies and business literature. As Duguid (2008) argues, its popularity may have been largely driven by the "seductive" term *community*, as well as by the choice of examples in the article that introduced the notion to the organizational literature (Brown and Duguid 1991), which inadver-tently helped management scholars take what was meant to be a critical perspective (rooted substantially in Marx's notion of "praxis") and incor-porate it into a standard toolkit for corporate "knowledge management."[1]

In addition to being partly responsible for this Panglossian turn, Lave and Wenger's (1991) use of the term "community" has influenced the evo-lution of the concept in another unfortunate way. The term "community" was generally understood as referring primarily to a group of people, often further imagined as a concrete, local, and often "tight-knit" group. In much of the recent literature, the term "communities of practice" is primarily used as a more positive synonym for workplace cliques.

If communities of practice are understood to be cliques, then Lave and Wenger's theory becomes largely reduced to the idea that knowledge and practice fundamentally reside in small groups and depend on face-to-face interaction. There is quite a bit of truth to this idea. As many authors have argued in response to the proclamations of death of distance, face-to-face interaction remains crucial for human relations. The success of the "infor-mation revolution" has not stopped the flux of talented engineers into Sili-con Valley but rather accelerated it (Brown and Duguid 2000). Nonetheless, overstressing the importance of local cliques leads us to a view of knowl-edge and practice that simply cannot square with the daily globalization as experienced by the people I met in Brazil. Driving through Silicon Valley along El Camino Real, one might sometimes forget about the world behind the two hill ranges that fence off the valley. Looking at the world from a place like Rio, however, the limitations of this view become quite clear.

When I started interviewing software developers in Brazil in 2005, they often asked me if I was a developer myself. When they did this, they were not inquiring whether I was a member of their local clique—they usually knew that I was not. Instead, they wanted to know whether I belonged to a large and somewhat abstract collective of people—several million around the world—who write software code. As a foreign member of this group, I was not expected to understand all the local meanings and norms. For instance, my interviewees took time to explain to me the many difficulties of doing software work in Brazil. They also pointed out specific people in specific organizations that I should talk to—again, correctly assuming that I would not know by myself who the important people were. At the same time, they expected me to understand their technical jargon (at least when used in English), as well as many of their values and practices. For example, having identified myself as a "former software developer," I was expected not only to know what a "source control system" is, but also to understand the technical and the social implications of the statement that a particular company lacked one. (At the technical level I would need to imagine the likely outcomes, while at the social level I would be expected to form the appropriate opinion of the people who run the company.) In fact, I often needed to make a special effort to make my Brazilian interviewees *suspend* the assumption that I share their meanings and opinions and to explain everything to me, *as if I were not one of them.*

This willingness to assume that I would understand their terms and practices is not a matter of wishful thinking. Brazilian software developers' work is, in fact, quite similar to that of software developers in other countries. As I will illustrate in more detail later, this involves not just facing similar problems, but solving them in similar ways, relying on the same set of concepts, calling relevant objects by the same names (either in English or by using Portuguese terms borrowed from English), and making many of the same jokes along the way.

Brown and Duguid (2000, 2001) attempt to address the limitations of such a "local" understanding of communities of practice by introducing the notion of "networks of practice," which they define as widely distributed networks in which local communities serve as nodes. People participating in such networks, they argue, can share knowledge across great distances by exchanging documents and other forms of knowledge media. As Duguid (2005) points out, however, successful communication in such networks *assumes* prior commonality of practice. Globally shared documents, representing the global "knowing that" (Ryle 1949), can be powerful because they go through local interpretation aided by the situated "knowing how" that the participants acquire through in-person participation in the local community of practice. We are thus left with the question of how members of remote communities come to share the situated knowledge that is said to be necessary for understanding globally circulating documents.

One possible answer is that practice is spread and synchronized by itinerant practitioners. The travel of people engaged in knowledge-intensive work has been noted by many authors (e.g., Traweek 1992; Castells 2000; Knorr Cetina 1999; Xiang 2006), and Saxenian (2006) puts circular migration at the center of her explanation of the development of links between the silicon chip industries of California and Taiwan and between the software industries of California and India. Despite the importance of such physical movement of people for certain kinds of practice, however, long-range migrants are still a fraction of the world's population and their travels tend to link rather specific locations (e.g., San Francisco and Bangalore). No matter how close San Francisco may be to Bangalore, most cities in the world, even large ones like Rio de Janeiro, are far from both in the experience of most people who live and work there. At least in the case of the software developers working in Rio de Janeiro, it is clear that most of the work of keeping the local practice "up-to-date" is done by people who rarely (if ever) leave Brazil. And while such people are hardly Castellsian "networkers" (the powerful globe-trotting "information brokers," according to Castells 2000), we also cannot dismiss them as the downtrodden servants of global capitalism ("the networked" in Castells).

Understanding those larger global collectives of practitioners as "networks" also places emphasis on the individual ties between practitioners, downplaying the importance of the *collective* nature of practice—a problem that plagues most attempts to model society in terms of "networks." Being a software developer is not just a matter of establishing ties with a few other people who engage in the same practice and then using those ties to pump information and influence. Rather, it involves identification with a named global collective ("developers," "coders," "pessoal do software") and acceptance of certain meanings and norms as meanings and norms *of the collective* rather than of specific individual practitioners. For this reason, I avoid the term "network" and instead describe those larger collectives as "worlds of practice."

I borrow the term "world" (or "social world") from the Chicago school of sociology, where it has been employed since the 1930s, most famously by Strauss (1978, 1982) and Becker (1982; see also Becker and Pessin 2006). This term has often been used to denote loose collections of people united by interests, outlook (Shibutani 1955), or activities. Social worlds can be quite large in their spatial dispersion and (unlike most notions of "community")

do not carry the implication that the members know each other or inter-
act on a regular basis. Strauss (1978, 1979) considers this dispersion quite
explicitly, describing the activities that define each world as occurring in
specific "sites", introducing the possibility of an analysis of the relationship
between the sites.[2] Levine's (1972) study of the art world of Chicago that
was quoted earlier in the chapter can be seen as an example of this kind of
analysis, pointing out the dependence of the art activities taking place in
Chicago on what is happening in New York City.[3]

Levine's discussion of the nature of this dependence also illustrates
Strauss's notions of "authenticity" and "authentication" in social worlds
(Strauss 1978, 1982). Unlike "networks," social worlds are usually understood as bounded, with a distinction drawn between members and non-
members. Membership, however, can be a matter of degree. As Strauss
(1978) argues, some participants may be seen as more authentic representatives of a particular world. This observation, he points out, raises questions
about the processes of *authentication*: "Who has the 'power' to authenti-
cate? and how? and why?" (123). Levine's analysis suggests that such power
may have much to do with place. Some places serve as "meccas" of their
worlds, carrying tremendous power not just in terms of practical ability
to coordinate resources but also as sources of legitimation and arbiters of
membership. This connection between space and authentication comes up
repeatedly throughout this book. To paraphrase Levine's interviewee, if you
want your software to be used widely in Brazil, you should write it in Silicon
Valley.

The concept of social worlds presents a number of challenges, in particular due to its substantial vagueness and its ambivalence about the locus
of agency in social worlds.[4] In developing my notion of "worlds of practice," I attempt to solve some of those challenges by building on a theoretical foundation provided by the idea of *practice*. In the next two sections I
explore the concept of "practice" in more detail, showing in particular how
it can give proper recognition to the agency of the people who participate
in such worlds while giving us a way of looking at structural constraints
on individual agency. This will help us understand how a world of practice
is gradually established in a new place and how it operates in a peripheral
location after that.

Practice and Structuration

Practice is a complex concept with a sinuous history. It is often employed
in modern social science without being defined and in ways that alternate

between its technical and vernacular meanings. A quick review of the concept's history may help reduce some of the mystique that is sometimes associated with practice theory.

In philosophy, the concept of "practice" goes back to Aristotle, who used the Greek term "práxis" (πρᾶξις) in two different ways. In some cases, he employed its vernacular Greek meaning of "action," which could refer to a wide range of human activity.[5] (In this way, the vernacular meaning of "praxis" in ancient Greek appears to correspond substantially, though not fully, to the vernacular use of "practice" in modern English.) However, Aristotle also introduced an additional technical sense of "praxis" to denote a particular kind of human action, which he distinguished on the one hand from "theory" (*theoria*, contemplation of the world aiming at truth) and on the other hand from "making" (*poiesis*, action aiming at creation of tangible products). This narrow sense of *praxis* thus referred to human activity that is transformative, yet free from economic rationale (and hence distinct from "work"). In later European philosophy, the distinction between *praxis* and *poiesis* largely disappeared, as the Enlightenment did away with Aristotelian prejudice against productive "making." (If the three-way distinction had been preserved, modern "theories of practice" would perhaps be more appropriately called "theories of making.") The remaining two-way dichotomy between practice/making and theory became increasingly important in Western thought, however, especially with the rise of the Industrial Revolution and Adam Smith's analysis of the role of labor in social life. A number of thinkers attempted to reconcile this newly discovered importance of practical activity with idealist philosophy (and in particular with German idealism). In particular, this task was undertaken by Karl Marx, who called for materialist philosophers to shift their attention from trying to understand how people theorize the world to looking at social life as "essentially practical" (Marx [1845a] 1978, 144). This meant, in particular, situating human consciousness in the context of "material practice," seeing today's world as a product of the historical work of previous generations (Marx [1845b] 1978, 170), and trying to understand social change by looking at human work to transform the world rather than at the evolution of ideas.

In this Marxian sense of "material practice," nearly all modern social theory is a theory of practice (or, perhaps, theory of poiesis). Even social theorists that focus substantially on the ideational aspects of social life typically consider those together with an analysis of members' practical concerns—for example, finding and holding a job or some other source of income. (Weber's classic *Protestant Ethic and the Spirit of Capitalism* is an example of this approach.) In the second half of the twentieth century, however, the term "practice" came to be associated with a more specific—though still quite diverse—set of ideas. In sociology, "practice theory" in this narrower sense has been exemplified, for example, by the work of Giddens (1979, 1984) and Bourdieu (1977).[6] Additionally, the work of many authors on whom I draw in this book (in particular, Latour and Becker) shares certain elements—different ones—of "practice theory." A related set of ideas has been identified with the philosophical work of Wittgenstein (see Giddens 1979; Schatzki 1996) and with the psychology of Vygotsky (1978, [1930] 2002) and his followers (Leontiev [1972] 1981).

One of the things that practice theorists share (to some extent, at least) is a particular approach to thinking about the relationship between social structure and human agency, which distinguishes them from two other approaches that are common in contemporary social thought. One of those approaches seeks to understand society by breaking it up into constituent units, such as countries, classes, racial groups, or corporations, and then analyzing the relations between such units. This "structural" approach assumes that a society can be meaningfully broken up in such a way and, further, that the relations between such units are more important for our understanding of the society than the specific interactions between the individuals. In other words, under this view, the society moves according to the logic of the units, not of individual people. A different approach stresses individual agency and microinteractions, putting into the background the larger social structures, if not outright denying their existence. In contrast to those two approaches, adherents of practice theory typically look to consider explicitly the relationship between agency and structure, refusing to give priority to one over the other. In most cases, they recognize that human action is strongly dependent on the structure of the social systems in which it occurs. On the other hand, they see such systems not in terms of an interaction of fixed units that follows its own higher-level logic, but rather as arising from the daily action of individual people and re-created in such action. Following practice theory typically means paying close attention to how social systems are constructed *and reconstructed* through individual activity and, at the same time, how they constrain individual action.

Anthony Giddens's theory of structuration provides one particular approach to such reconciliation between agency and structure. Giddens argues that the notions of action and structure presuppose one another. For Giddens (1979), structure has "virtual existence" (63), being real in its effect on human action yet existing only in the social interactions. Social systems, Giddens argues, "cease to be when they cease to function" (61). Giddens offers an analogy using language: grammar rules structure the speech of individual speakers, yet such rules persist only as long as they are reproduced by the individuals in the numerous acts of speaking.

The focus on the interaction between agency and structure leads adherents of practice theory to view society as composed not just of units that are capable of making decisions (e.g., nation-states and corporations) or groups defined by specific characteristics of the members (e.g., by socioeconomic status), but rather of "practices"—reproduced patterns of human activity or "regularized acts" (Giddens 1979, 56). In Giddens's analysis, practices are *co-constitutive* with social structure, which itself becomes understood not as a set of entities, but rather as the regularity of the actions of the individual members.

Giddens uses two concepts to explain the process through which the regularity of social systems is reproduced. The first is members' knowledge of how their social system works and how things are normally done. Following Sewell (1992), I call such understandings *schemas*.[7] (Sewell's term also points to parallels with the notion of "schemas" in cognitive psychology, though it is important to not overstate this parallel.[8]) Schemas help maintain practices by allowing individual actors to recognize the regularity of activity and to proceed in accordance with this recognition—usually in a manner that ensures that the regularity is maintained. For example, when we enter a restaurant, we recognize this social environment as "a restaurant" and proceed to act as customers, because we know that doing so will get us the results we want.

While individuals can have their own idiosyncratic schemas, the schemas are particularly powerful when they are shared. Shared schemas allow actors to set expectations for what others would do. When we invoke the restaurant schema upon entering a restaurant, we do this with an expectation that other members of this social setting will act out *their* roles. It is this expectation that makes it sensible for us to act out our role as customers as defined by the schema. Sewell points out that collective schemas correspond to what anthropologists call "culture." I similarly use the word "culture" in this book to refer to a set of schemas held collectively by a group of people.

Practices are also reproduced through use of *resources*. To borrow Sewell's "translation" of Giddens's somewhat cryptic definition, resources are anything that people can use as a source of power (Sewell 1992, 9), which in particular can include material objects, as well as human energy and skills. In Sewell's reformulation of Giddens's theory, schemas and resources are the two main components of social structure that remain in a duality, simultaneously requiring and sustaining each other. Actors configure material and human resources in accordance with their schemas. Those configurations of resources then give schemas a tangible form. The actors can also "read" configurations of resources to recover schemas from them (13). Such configurations can of course also provide a context that helps the actors recover schemas from actual texts.

Applied to software development in Brazil, this perspective would lead us to look not at the interaction between Brazilian state agencies and software companies, or at the subjective experiences of the individual programmers, but rather to put at the center the different "doings and sayings" (Schatzki 1996) that are involved in the development of software. In analyzing such doings and sayings, we would pay attention to the actors' use of schemas and resources, asking how individual software developers gain access to resources, how they configure and interpret them, and the different sources on which they draw for their knowledge not only of software itself, but also of the social structure of the software practice. This, in turn, opens up the possibility of a more careful analysis of the nature of ties between software practice in different places.

Once knowledge is understood as rooted in a *practice* rather than merely in a *community*, its mobility becomes dependent not on the limits of group interaction but on the spatial dimensions of the practice itself. We must therefore understand not just how words or people reach faraway places, but how a system of doings and sayings becomes established there. Words can be brought to new places and yet not be understood. People can move and proceed to doing something else. As a system of doings and sayings is established, however, it can draw in new people (including those who have never traveled) and provide a fitting context for words and things arriving later.

Imagined Practice

To understand the process through which a practice is established in new places, we must consider some of the sources of *change* in practices. While systems of activities maintain continuity through the mutually sustaining relationship between schemas and resources, this continuity should not to be interpreted deterministically. While some theorists of practice (e.g., Bourdieu) see practices as almost self-reproducing, I follow Giddens and Sewell in stressing that practices *are reproduced by actors*, who possess real agency and often make real choices, at least within the constraints of resources available to them. Schemas are not programs run by human computers, but rather models of the world that help individuals decide how to proceed in a given situation. Individuals may find that multiple schemas fit a situation, and they may choose between them. They can "transpose" schemas from one situation to another. They can reinterpret resources, finding new schemas in them. Finally, as members of a society come to know more about their society (through social science research, among other things), they develop new schemas, which leads them to the construction of new practices.[9]

Bringing a practice to a new place is a form of change, involving reconfiguration of material resources in the new place as well as a change in schemas used by people who live there. This change, however, is not merely a failure of continuity. Rather, it is often driven by purposeful activity of people who seek to establish the new practice in this place. We can think of such active expansion of practice as itself *a kind of structuration*. In traditional structuration (Giddens 1979), the members reproduce social structure through their actions that are guided by their knowledge of the existing social structure, resulting (for the most part) in reproduction of the existing patterns of interaction. In the "long-distance" structuration, on the other hand, local actors organize their actions relying not on their knowledge of their own society and their own current practices, but rather on reflexive representations of *remote* social systems.[10] The result is often a change in the local system of interactions that brings it closer to the remote system.

The use of remote schemas involved in this kind of structuration is a particular case of what Sewell (1992) calls "transposability" of schemas: the actors' ability to draw on schemas that were originally developed for use in a different context. We should note, however, that the schemas that are being transposed may in this case be acquired not through lived experience in the remote social systems, but from *descriptions* of such systems. This means that we must investigate how local practitioners achieve their understanding of the remote practice and social structure, considering among other things the specific proactive uses of communication technologies to learn more about what is happening outside. While practice theory suggests that reproducing a practice based on blogs is an uphill battle, we should not assume that such reproduction does not take place. We must also, however, be mindful of the potential gaps between peripheral members' models and reality, and consider the ways in which they may identify and mend such gaps.

While local members may know a good deal about foreign practices and social structure, this does not mean that they always draw on this knowledge as a structuring resource. The argument that a California company does things in a particular way may carry a lot of weight in some situations, but it may also be dismissed with a reminder that "we are not in California." This has much to do with local members' understanding of the ways in which the foreign context differs from their own, which in turn depends on their model of the world as a whole and their place in it. For example, their view of themselves as living in "a developing country" becomes crucial in negotiations of when the imported schemas apply.

Structuration over distance often requires a substantially larger leap of faith. When a social system maintains continuity, this happens because the members know that it most probably will. Structuration over distance, however, requires that activities are structured by a collective understanding not of what *is* (or will be) but of what *is possible*. (This is to some extent true of most social change.) Appadurai's (1996) discussion of "imagination" and "the imaginary" ("a constructed landscape of collective aspirations," 31) can help us understand how reflected understanding of foreign structure is used in structuring local action. Reflected foreign practices and structure provide local actors with the elements for constructing imaginary worlds in which there would a place for them. Once articulated and shared, such imaginary worlds can become blueprints for action.[11]

Imagination can also inhibit action. A Brazilian proverb says that a dog once bitten by a snake becomes afraid of a sausage. When looking at Brazilian software development today we must consider, in particular, the ways in which one such past snake accident (the "failed" attempt to build a sustainable computer industry in 1970s and 1980s, discussed in chapter 4) affects today's actions.

Appadurai (1996) draws a distinction between "imagination" and "fantasy," seeing the latter as private and as carrying "the inescapable connotation of thought divorced from projects and actions," contrasting it with the *projective* and *collective* qualities of "imagination" (7). I believe, however, that we must recognize the fuzzy boundary between the two and the importance of what I call "subvocal imagination": imagined worlds that are too unlikely to be publicly presented as a serious basis for actions, but that nonetheless influence action in profound ways. I look at this issue more closely in chapters 6 and 8.

Bounding Practice

I started my discussion of "communities of practice" by pointing out the danger associated with the term "community." I then unpacked some of the theoretical apparatus represented by the term "practice," showing how theories of practice can help us understand the reproduction of social systems across space. By putting aside the term "communities," however, we left out of sight a few important aspects of practice. Systems of activities such as software development are often bounded, with an important distinction drawn between members and non-members. It is for this reason that I call such systems *worlds* of practice, using the term "worlds" to connote collectives of a particular kind—ones that cannot be called "communities" in the full sense of the word, but which share important aspects with communities.

Software developers, whether in Rio or in Silicon Valley, often participate in small groups of fellow-practitioners which can be called "communities" in the full sense of the word—that is, possessing all characteristics that can be expected of a "community." Such characteristics would include physical proximity of the members, a dense pattern of interactions between them, and collective identity—namely, the members' recognition of the group as a meaningful unit. Access to such communities is crucial for successful participation in many activities. Perhaps one of the main reasons for this is that a combination of shared destiny (arising from the shared location) and past history of interactions facilitates trust. Within a community, one can know whether a particular person has betrayed trust before and can be assured that a future betrayal would be costly. Collective identity provides additional mechanism for identification of members. This increased level of trust can facilitate joined projects tremendously. In chapter 5, for example, I describe how being identified by Rodrigo as "a Herculoid," a member of Rodrigo's community of tech-minded professionals, gained me access to Alta, a software company on which that chapter is based.

Practitioners can also participate in communities that are characterized by interaction and collective identity, but not by geographic proximity. For example, in the course of my work I have also become a member of what my interviewees call "the Lua community"—that is the community of people formed around *lua-l*, Lua's main mailing list. Having been subscribed to *lua-l* for many years, I have learned to recognize many of the names and have been myself acknowledged as "a regular." While such communities rarely replicate the richness of dense face-to-face interactions, they may be strengthened with occasional face-to-face encounters.[12]

While the two kinds of communities described above are important for understanding practice (and I try to bring them in focus repeatedly in later chapters), we often cannot properly understand them without placing them in the context of yet larger collectives. In the case of the software developers whose experiences I describe in subsequent chapters, we must recognize that in addition to some of them being members of "the Herculoids" and "the Lua community," they also are a part of the global collective of people working with software. This global collective does not have the dense pattern of interactions that characterize smaller communities, but its members do share certain things, including a collective identity, expressed through overlapping labels such as "developers," "coders," "computer nerds," and "geeks."[13] If the smaller communities can be understood by analogy with a village, such larger collectives can be compared to a nation. One can rarely understand a modern village without considering the nation-state of which it is a part—both in reality and in the member's understanding of this reality. Similarly, one cannot hope to understand "the Lua community" without considering critically the notion of "software developers." When such larger named groups are organized around practices, I refer to them as "worlds of practice."

Like nations in Anderson's (1991) analysis, worlds of practice can be understood as "imagined communities." To appropriate Anderson's quote, the members "will never know most of their fellow-members, meet them, or even hear of them, yet in the minds of each lives the image of their communion" (6). This term "imagined" should not suggest that such "communities" are not *real*. Like nations, worlds of practice are also tied together with actual similarity of practice, joint actions, and circulation of ideas and material resources. As I will try to show, however, this similarity, joint actions and circulation are themselves to a large extent supported by the shared sense of communion. This joint re-enforcement, is, of course, fully compatible with the larger theme of structuration if we recognize that a collective identity and the notion of "community" are themselves schemas, which exist in a mutually sustaining relationship with configurations of resources (in Sewell's formulation) and with practices (in Giddens's formulation).

While imagined communities can span expansive distances, they must nonetheless be bounded. Members of an imagined community conceive of themselves as belonging to a *particular* group of people—one that does not include everyone. One cannot be a member of a kind without there being others who are *not*. Being able to draw a boundary between members and non-members is essential for forming a community. Such boundaries might not always be clear and may need to be negotiated, yet they must exist at least in principle. And the more interaction there is between members and non-members, the more carefully the boundaries may have to be drawn.

The existence of boundaries is important for several reasons. First, it facilitates collective action. Members can be expected to possess certain schemas and to apply them. Members who fail to act in accordance with the appropriate schemas can be sanctioned. Once two people identify themselves to each other as "software developers," they establish common ground, which facilitates joint projects and conversation. Second, maintaining boundaries between members and non-members allows for collective control of resources. Certain resources cannot be easily held by individuals but also cannot be left to be spent by anyone. Collective control by a group with defined boundaries provides a solution.

I can illustrate and expand those notions by drawing on Hughes's sociology of occupations. In his book *Men and their Work*, Hughes (1958) argued that occupations are defined on the one hand by "culture and technique" and on the other hand by a "mandate." Members of an occupation possess a particular "technique"—a set of methods for manipulating relevant objects. They also share a "culture" ("a set of collective representations"; cf. Sewell's "schemas") which allows them to see the world in a particular way—differently from how non-members see it. Perhaps the most important aspect of occupational culture is that it leads to a particular view of the activities in which the members engage, of why they engage in those activities, and of what sort of people they are. As we will see in chapter 3, for example, software developers collectively see software development as an intellectually challenging line of work and as something that ought to be done out of passion for technology rather than just as a way of earning money.

While the technique and culture can be seen as collective resources, few modern occupations (or worlds of practice more broadly) attempt to exercise active collective control of their culture and technique by preventing outsiders from learning about them. Rather, the most important collective resource is the other defining element of occupations noted by Hughes (1958)—what he calls a "mandate." To understand Hughes's notion of mandate, we need to first consider that practices like "software development" do not exist in isolation. Rather, they form a part of *a system of practices*, of societal division of labor.

Software developers often describe programming computers as intellectually stimulating and "fun," talk about being "in love" with software, and expect others developers to see it in the same way. We must not forget, however, that, ultimately, people who program computers usually do so because other people have a need for programmed computers. For most of the people whom we will meet in this book, writing software is a job that brings income. This is, of course, no accident. People who program computers can only do so if they have some way of satisfying the basic necessities of life. After all, the time spent programming a computer is time not spent farming, hunting, or making clothes. The practice of software development also requires access to certain equipment, such as computers, for example. Making a computer in turn requires not only knowledge that software developers usually do not possess but also millions of dollars' worth of *other* equipment, which itself in turn needs to be manufactured. In other words, the practice of software development requires material resources that must be supplied by *outsiders*. This practice is made sustainable because engaging in it can produce something that is of interest to non-members, supporting *other* practices. This means that a world of practice cannot be fully understood (and perhaps cannot even be meaningfully defined) outside of its relationships with other worlds and with non-members.

Dependence on resources controlled by outsiders makes a practice dependent on *schemas* held by such outsiders. Such schemas define "mandates" assigning a practice a legitimate place among other practices and recognizing that some people are allowed (or even expected) to engage in it. Mandates are collective—instead of recognizing that specific individual people can legitimately engage in particular activities (say, establishing that Mr. John Smith can legitimately stick needles into other people's bodies), they define categories of people (e.g., "doctors") who can engage in such activities, and even get rewarded for this. Similarly, a manager of a retail company looking to build an e-commerce system would look for "a software developer" (and not, e.g., "a lawyer") to build such a system. A particular individual must typically be recognized as a member of that category to get his or her individual "license." Further, what often matters is being recognized as a member of the category *by outsiders*.

In some cases, the mandate is formalized—certain groups are given a legal monopoly to perform certain tasks. For example, in the United States, the work of selling securities is by law reserved for a particular kind of people, who are colloquially called "brokers" and more formally referred to as "registered representatives." In sociology, such groups are called "professions." When the mandate is formal, the boundaries of the group are usually also quite formal. In the United States, securities representatives must be registered, having passed certain exams, administered by an agency acting under government oversight.

Software developers do not form a profession. Neither in the United States nor in Brazil is there a legal limitation on who can write software. Nonetheless, the notion of "mandate" applies to software development as well: there is still a general understanding that certain tasks should be done by "software developers." The boundaries of this latter category are informal, but this does not mean that they do not exist. The outsiders need to know who can claim the mandate that has been given to the group, which means that claims to membership must be somehow evaluated. The methods of authentication are themselves subject to negotiation. In particular, the insiders must do the work of straightening the boundary between members and nonmembers by establishing the "proper" authentication procedures and educating outsiders about whom to properly recognize as true members. Gieryn (1983) calls this active process "boundary-work," which can be understood as the work of crafting and disseminating schemas that insiders and outsiders could use for authentication of members.[14]

While this approach suggests that group boundaries can be a useful tool for maintaining control of resources, it does not mean that the members necessarily see this process in such a way. In fact, the boundaries can often be enforced most effectively when the schemas that govern authentication naturalize such boundaries. National communities, for example, are often understood by their members as "primordial." While modern sociology typically finds that "nations" as we know them today are recent and may be best understood as "constructed" over the last few centuries, we cannot deny the subjective reality of eternal national communities. We must thus recognize the "constructed primordialism" of nations and ethnic groups (Appadurai 1996). In the same way, we must take into account computer nerds' view of themselves as a different kind of people, perhaps even born with different brains, while at the same time considering how such understandings of who they are may be implicated in boundary work and in the maintenance of class divisions.

The boundaries of worlds of practice are particularly important for understanding reproduction and expansion. Becoming a practitioner means crossing the boundary, turning from an outsider into a member. In some cases, this process is straightforward and can be understood through what Lave and Wenger (1991) call "legitimate peripheral participation": a novice joins a community by following the steps that the community has accepted as proper ways of joining it, perhaps taking a subordinate role to the older members.[15] Through active interaction with members, the novice learns the culture and technique, and becomes accepted as a bona fide member, transitioning to the more central forms of participation.[16] Through such acceptance a novice gains access to resources controlled by the community and can use the community's mandate to access the relevant resources controlled by outsiders. Such access will become crucial for newcomers who decide to stay for the long term. This, however, may not necessarily be their goal at the beginning. They may join the world of

practice as *a community* without much interest in the actual practice, especially in its economic dimensions. As I show in chapter 3, early engagement
with software can simply be a way of "hanging around" with fellow nerds.[17]

What happens, however, when someone attempts to join a world of
practice in a place where there is no local community of practitioners—or
where local communities that *claim* to represent the world may themselves
have a hard time convincing others that they can do so successfully? To
understand this, I contend, we must again remember that worlds of practice do not exist in isolation. They exist as an interdependent system, and
they expand *together*. The practice of software development, for example,
did not arrive in Brazil by itself. Rather, throughout the twentieth century,
many American practices were being re-created in Brazil. As some remote
practices were re-created, people who engaged in them needed the support
of the related practices. In the case of Brazil, for example, we will see how,
among other things, the Brazilian government's desire to keep its governance practices in sync with foreign models required synchronizing the
practice of census-taking. Taking a census the American way required using
a computer. Once a computer was brought, someone had to program it.
The practice of programming was therefore not reproduced in a vacuum,
but rather came together as a part of a larger process of "modernization,"
as I show in chapter 4.

This process of parallel re-creation provides prospective members with
some of the key resources they need to start the reproduction of a practice. It often does not provide them with everything they need, however.
Full membership in a world of practice requires applying proper culture
and technique while engaging with proper resources, obtained through a
successful claim to the group's mandate. Those two sides of the practice—
the cultural and the material—must be mutually sustaining. The culture is
learned through engagement with resources while resources are acquired
through demonstration of proper culture and technique. It is important to
realize, however, that those two sides of the practice do not represent its
essential qualities, but rather function as *discursive tools* that can be used by
members and nonmembers to negotiate rights to engage in certain practices or to call what they do by certain names. Fulfilling the role de facto
is one test individuals can try to *use*. The ability to demonstrate similarity
to other members is another test. When those tests yield different results,
the individuals may engage in negotiation as to which one is more relevant
in a given context. Local "boundary work" can further be used to educate
the local public about what categories should be considered important in
specific circumstances. In other words, we can look at the different kinds

of "moves" that participants can make, using resources that they have to obtain those that they lack, increasingly establishing themselves as bona fide members of the practice.[18]

Practice at the Periphery

The process of reproduction through expansion that I described earlier does not usually result in a global world of practice spread evenly among all of its sites. Rather, the sites continue to vary in power and significance. Some function as centers of the world, defining the practice and coordinating global activities. At such "centers," the local community's authenticity is rarely questioned and it suffices for an individual to focus on finding his or her place in the local community, without needing to worry where this community fits in the larger world. This symbolic capital possessed by the central actors goes together with control over the much more mundane forms of capital. As I point out in chapter 4, two metropolitan regions host headquarters of firms that jointly control over 60 percent of world's software industry capitalization. Work done elsewhere is often directly controlled from those places.[19]

Other sites are *peripheral*.[20] From the central perspective, they are recognized as present but unimportant. From the perspectives of local outsiders, the local practitioners may be "good enough" as providers of services, but their status can be questioned. This has important consequences for how practice proceeds at the periphery even after it is established there. This book concerns itself with the periphery.

When looking at a world of practice from a central location, it is sometimes possible to get the impression that everything that matters happens right there. Peripheral practitioners, however, can rarely afford such a limited view. They are judged—collectively and individually—on their ability to represent the global practice, to solve local problems not with local solutions but as central members would have done—for example, developing a software solution as it would be done "over there in Silicon Valley." The need to be recognized as authentic representatives serves as an important structuring (and synchronizing) resource, because it gives local members (and sometimes outsiders) the ability to censure lack of compliance with the "standard" practice.

The symbolic value associated with remote prototypes often presents peripheral practitioners with a difficult choice: they must decide when to cast their lot with the local community and when to seek direct ties with those parts of the social world that lie outside. To understand what happens at the periphery, we therefore must often consider side by side at least three entities—the individual, the local community (with all of its factions), and the larger world with its central sites—noting the different ways in which the individuals may attempt to "escape" the local community by establishing direct links with the remote centers. By understanding what drives the individuals to build such ties, we will come to understand not only how the local community synchronizes its practice with the rest of the world, but also why peripheral communities may often face problems organizing for collective action.

Peripheral participation in a practice may often involve a more complicated interaction with the broader local society for several other reasons. Over time strong worlds can transform the society around their centers, making it easier for the members to move back and forth between their world of practice and the mainstream society. For example, while the "nerd" identity associated with software developers was seen in United States as somewhat "unmanly" in the past, it was partly incorporated into "hegemonic masculinity" during the 1980s and 1990s (Kendall 1999). Due to this historic work, a software developer working in Silicon Valley today rarely experiences conflict between his identities as "a man" and as "a software developer." Developers working in Rio de Janeiro, on the other hand, operate in a place where somewhat different forms of masculinity are the norm. Reconciling the identities of man and nerd is thus harder in Rio than it is in San Francisco.

A world of practice may also be at peace with the mainstream culture at the center because aspects of that mainstream culture are often incorporated as basic assumptions into the culture of this world, and even into its material artifacts. Peripheral members, on the other hand, again have to face the contradictions between the demands of the world of practice brought from abroad and those of their local mainstream society. This effect is easiest to see with language: software developers in California can perform all of their daily activities in one language, while their Brazilian counterparts must switch between the language of the software world (English) and the language of the local society (Portuguese). The choice of language in specific contexts can thus serve as an important marker of allegiance to either the local community or the global world of practice (see the next chapter).

The asymmetric relationship between the center and the periphery and their different relation to their respective local societies have important consequences for the flow of innovation. New practices and knowledge produced at the center are often mobile from birth. While such practices and knowledge may be inextricably tied to local culture and context, this is not an unsurmountable problem, since the rest of the world of practice is typically ready to accept such practices on those terms. A book on software development written in California in English does not need to be translated to become successful worldwide: the author can count on the potential readers either learning English, the language of Silicon Valley, or struggling through the book with a dictionary. Practices and knowledge generated at the periphery, on the other hand, have little chance of success outside their local context unless they are actively disconnected from it. In other words, central actors can "disembed" their knowledge using the simplest strategy available, leaving others the hard work of reembedding it at the periphery. Peripheral actors, on the other hand, must perform the most thorough disembedding, to make reembedding at the center a trivial task. In doing so, they may have to forgo the needs of local users (or at least the less globalized ones), as we will see in the case of the Lua programming language in the next chapter (and again in chapters 6 and 7).

In doing so, peripheral participants help re-create the asymmetries from which they suffer themselves. Latour (1987) argues that the foundation of European science lies in the massive accumulation of basic knowledge of the world made possible by Europe's central position in a system of colonial empires—a place where knowledge and resources were brought from around the world. In addition to peripheral plants, animals, and cultural artifacts brought for examination to the centers, this accumulation often involved peripheral individuals themselves. While a few centuries ago colonial subjects were often brought to the center by force, today many go there of their own will. Often, it is the most talented of the peripheral individuals who gather at the center. While their move to the center may strengthen the integration of the peripheral site with the center, it often also leaves behind broken alliances. Those who remain at the periphery, however, also contribute to the reproduction of central power, often dedicating their work to bridging the remaining gaps between the local context and the resources deployed from the centers. In doing so, they often make such resources (and those who control them) even more powerful.

When considering the persistent differences between the centers and the periphery, we must also take into the account the factor known as "network effects," or, more formally, "network externalities." In economics, the term "network externality" denotes the additional value that is enjoyed by users of a particular technology when more people come to use it. The concept originated in the context of telephony when it was observed that each new user who signed up for telephone service made the telephone system more valuable for other subscribers who gained an additional person whom they could call.[21] Network externalities can also arise because additional users stimulate the supply of complementary products by increasing the economies of scale for the production of such goods. For example, Katz and Shapiro's (1986) classic paper on network externalities discusses the adoption of VHS VCRs: more people buying VHS VCRs led to the higher availability of VHS video rentals, which in turn made VHS VCRs more valuable. Network externalities can therefore lead to "the rich get richer" effect where technology that has been adopted by many people becomes increasingly adopted.

Software products and computing services are often characterized by especially strong network effects. For example, the wide use of the Java programming language brings numerous benefits to people and companies who use it. Java programmers can benefit from Java modules and frameworks written by other Java programmers. They also have access to a variety of jobs that require Java expertise. The companies, in turn, have access to a variety of programmers.

While the concept of network effects originated in the context of technology adoption, it is important to realize that similar effects can happen around elements of practice that are not technological in the narrow sense of the term. For example, Grewal (2009) notes the network effects associated with the use of English as a lingua franca. Each additional person who chooses to use English adds to the value of the English "network," making use of English yet more advantageous. Network effects can also be associated with *places*. In many cases, network effects form around technologies and other elements of practice that are strongly linked with specific sites. For example, the Java programming language was developed in Silicon Valley and is today controlled by a Silicon Valley company. Network effects can also, however, take place around locations in a more direct way. For example, the more engineers and entrepreneurs move to Silicon Valley, the more attractive it becomes for engineers and entrepreneurs.[22]

While economic literature typically sees network effects as creating additional "value" associated with specific technologies, network effects can also be seen as oppressive in that they constrain individual choice. Grewal (2009), for example, argues that the wide use of English does not merely provide additional value to those who choose to speak it. Rather, in many contexts today it eliminates the choice altogether, making English the only viable option. (I demonstrate some of this dynamic when I discuss the use of English by Rio software developers in the next chapter.) Consequently, Grewal uses the term "network power" to refer to the constraining nature of network effects. In my own analysis I avoid such terminology since I aim to highlight individual agency. For example, instead of accepting the developer's choice of English in some context as simply a demonstration of the "network power" associated with the language, I focus closely on how participants' make their choices, even in cases where such choices seem obvious to the participants themselves. I also contrast cases where English is chosen to those where it is not.[23]

It is important to recognize that in the long term peripheral re-creation of practice can give rise to alternative centers. While most parts of the world are unlikely to ever occupy this role, *some* of them can rise to rival the earlier seats of power. Silicon Valley, after all, was not the place where computing or programming was originally developed and, in fact, had little more than citrus orchards just a bit over a half-century ago. More recently, Bangalore has become a new important site in the world of software. (Other cities in India, such as Chennai and Hyderabad, have also grown a substantial software industry, though they have received less press.) Bangalore is far from becoming a rival of Silicon Valley and at the moment plays a clearly subordinate and dependent role in the software world: the best work available to software developers in Bangalore today is provided by companies based in the United States, often headquartered in Silicon Valley. This reality highlights, however, the dynamic nature of the worlds of practice.

New centers, of course, often have a steep path ahead of them. Quite often, their success requires that the local members of the world of practice create an enclave, separating themselves successfully from the local context. Bangalore's Embassy Golf Links Business Park, home to some of the city's most prestigious IT employers, stands in striking contrast to the neighboring parts of the city—the city that itself is a world apart from the rest of the country. Bangalore's success might also be attributed to the city's relative lack of commitment to any local language and its willingness to adopt English as the working language. As we will see, Lua's success in its niche in many ways involves proper management of distance from local institutions.

Newcomers are often also handicapped by lack of proximity to the centers of *other worlds*. The different worlds of practice form an interdependent system. Consequently, their central sites coincide to a substantial degree. San Francisco is a center of a number of worlds, which reinforces its position in each of them. On the other hand, while Helsinki had become somewhat of a mecca of the world of mobile software development at the time I was doing my fieldwork in Rio, few of my interviewees were contemplating learning Finnish. It is perhaps not a big surprise, then, that Finland's leadership in this area started to fade in recent years, as the field became dominated by two new entrants from Silicon Valley. This means that while considering the geography of individual worlds of practice, we must keep in mind the politico-economic structure of the planet as a whole. Rio's position in the software world in many ways corresponds to Brazil's overall position in the world economy. The fortunes of Rio's software will likely continue to be influenced by this.

* * *

This chapter has outlined a theoretical framework for thinking about practice in space, developing the concept of *worlds of practice* and showing how we can think about such worlds as global yet at the same time tied to specific places. In particular, the concept of worlds of practice helps us recognize both the agency of individual people who do the work of expanding worlds of practice to new places and the structuring resources that are offered by the worlds themselves. The next seven chapters of the book illustrate more closely the different aspects of this model.

I start with a look at the use of English by Rio's developers, which helps illustrate in about the clearest form some of the ideas presented in this chapter. I then go back in time, looking first at personal histories of developers entering the world of software in chapter 3, and then at the larger history of software practice in Brazil in chapter 4. Chapters 5–8 look more closely at some of the specific contexts, illustrating the different configurations of local and global resources.

2 The Global Tongue

It was a cool day in June, one of the coldest months in Rio de Janeiro, yet the air conditioner was running on high, making me think that I should bring a sweater next time to avoid catching a cold. I was in the office of "Alta," a company in downtown Rio that, according to its promotional materials, focused on attending to the desires of its clients through the use of the newest technologies. In 2007 this was understood by many to mean building web applications in Java, and this is what Alta did. I was still getting to know the company and had just been set up with access to its intranet web site. I logged in and clicked around, browsing a number of pages: technical documentation, company policies, project descriptions. Everything was in Portuguese. The situation changed suddenly, however, when I arrived at the actual source code. Below is an example of what I saw:

```
/**
* @param weekday - Dia da semana em que o chronoEntry se encontra
* @param start - Data de inicio do Cronograma
* @param oldDate - Data do chronoEntry que esta sendo clonado
*/
private Date weekConvert(Integer weekday, Date start, Date oldDate){
Calendar cal = GregorianCalendar.getInstance(new Locale("pt_BR"));
cal.setTimeInMillis(0L);
//Data do inicio do Cronograma, assumimos que seja sempre segunda.
cal.setTime(start);
//Ajustamos o dia da Semana
cal.add(Calendar.DAY_OF_MONTH, weekday.intValue() - 1);

//Copiamos Hora e Minuto
cal.set(Calendar.HOUR_OF_DAY,oldDate.getHours());
cal.set(Calendar.MINUTE,oldDate.getMinutes());

cal.set(Calendar.SECOND,0);
cal.set(Calendar.MILLISECOND,0);

return cal.getTime();
}[1]
```

The code contained a mix of Portuguese and English—or, to be more precise, a mix of Portuguese and *Java* (though as we will see, the boundary between Java and English is often hard to delineate).

Such a mix of languages in Alta's code in many ways reflects the encounter of the two worlds: the local world of Rio de Janeiro and the global world of software practice, giving this encounter a rather concrete expression. As I found during my fieldwork, the activities and discourse around the choice of language in this and other contexts quite often express most clearly the contradictions inherent in the peripheral participation in a global technical practice. For this reason, I look here at the use of English across a range of contexts that I explored in my time in Rio, putting some flesh on the theoretical skeleton presented in the previous chapter.

The Language of Software

Linguists use the term "diglossia" to refer to the situation where a single social group routinely uses two different languages, with most speakers being relatively proficient in both. In a typical diglossic system, the two languages have different roles. One language, which linguists typically call "high" or "H," is used for formal communication as well as for high culture. Another, "low" or "L," is reserved for informal communication, especially among close friends and family. Use of the high language connotes professionalism, education, culture, status, hierarchy, and commitment to larger (national or international) institutions. Use of the low language connotes intimacy, equality, and commitment to the local place.[2]

It would not be accurate to say that Brazilian society, or even more specifically the community of software developers of Rio de Janeiro, is diglossic in its use of English if we use the word "diglossia" in its narrow linguistic sense.[3] While some Brazilians learn to speak English fluently, it is still a foreign language in Brazil. In my time in Rio, I have never heard two Brazilians speaking English to each other, except for the sake of a foreigner or as a joke. Apart from my conversations with Lua's authors, nearly all speech present in this and subsequent chapters is my translation from the original Portuguese. Written English, however, is omnipresent in the work of Rio's software developers, as are short spoken phrases, which may or may not be altered to comply with Portuguese phonology. While such coexistence of English and Portuguese is not diglossia in its traditional sense, it retains some of its features: the high language (written English) can be used to communicate and develop status and global links, while the low language (Portuguese) builds local connections.

While linguistic literature on diglossia often focuses on the mechanics of code-switching and second language acquisition, diglossia is nearly always a power-invested phenomenon, and the social side of diglossia often cannot be understood without considering how the two languages tie together local power relations and external resources. In a typical case, proficiency in "H" marks the individual's status vis-à-vis the less proficient speakers, becoming an instrument of exclusion. Such proficiency is often predictive of the individual's social status because it requires access to educational resources available to only the privileged few. H also becomes important for local power relationships by connecting proficient speakers to a powerful external community interacting in H, allowing them to draw on the resources of this community. At the same time, however, such use of H often underscores its speakers' dependence on external resources and their subordinate position vis-à-vis the group that defines the norms of H. Speakers of H may often prefer to use L as a way of marking their opposition to that group and their connection to the local community.

What has been said about the relationship between a "high" and a "low" language applies to the quasi-diglossic relationship between English and Portuguese among Rio's software professionals. Proficiency in English (and especially the ability to speak it fluently) often reflects a higher socioeconomic origin. At the same time, it gives developers access to crucial foreign resources, further elevating their status by helping them acquire cultural capital in the global world of software. It *also*, however, highlights their peripheral status in a largely foreign practice of software development. They often downplay such tensions: good software professionals are expected to display a global perspective on the world and to accept the dominance of English without any nationalistic qualms. At the same time, many of the interactions presented in this chapter show elements of resistance and the careful handling of misalignments between the local and the global nature of the practice.

This quasi-diglossic relationship between English and Portuguese is just one of the dimensions of the larger phenomenon that we can call "cultural diglossia": the situation where a particular social group is simultaneously engaged in two cultural systems, which stand in asymmetric relation to each other. In our case, it is, on the one hand, the system of joint activities localized in Rio de Janeiro (and more broadly in Brazil) and, on the other hand, a global world of practice centered far away. The social dynamics of cultural diglossia express themselves in subtle ways. The choice of language, however, is often much easier to observe and discuss. For this reason, a careful look at language is a good starting point for the broader discussion of the reproduction of a professional practice.

When looking at a developer's choice of language, however, we must remember that in many cases the individual developer has very limited choice in the matter. English is not only intertwined with the culture of software, but is also thoroughly embedded in its very *technology*. Software programs are machines built of words (Samuelson et al. 1994). Additionally, to achieve any nontrivial goals, a piece of software must work together with other software. This compatibility is achieved through the use of the right *words*—usually the right *English* words. Consequently, "programming in Portuguese" often becomes either cumbersome or outright impossible. For many of my interviewees, the idea of using Portuguese for writing programs falls somewhere in the range from "ugly" to "ridiculous" or even simply unimaginable.[4]

To understand the extent to which English is intertwined with software, let us have another look at the software snippet shown above. The snippet contains two kinds of text. The text enclosed in "/* . . . */" (the first five lines) and the three lines that start with "//" are *comments*. Such text is ignored by the computer and is added solely to assist a human reader of the program. Such text can be written in any language, though the programmer is usually limited to the twenty-six-letter version of the Latin alphabet as it is used in English.[5] In the previous sample, the developer wrote all the comments in Portuguese, although he had to forgo accents in words like "início." (I use "he" since all of Alta's developers at the time were men.) The remaining text, however, contains instructions for the Java compiler, a program that converts those human-readable instructions into a much longer sequence of more detailed instructions that can then be executed by the computer. The developer had a lot less choice here—the instructions must be written in such a way that the compiler would be able to understand them. (Such instructions are meant to *also* be read by human programmers who will be revising the program later, but the compiler is their primary "audience.")

Three of the two dozen English words used in the instructions ("new," "private," and "return") are *keywords* (or *reserved words*) of the Java programming language. Together with fifty-six other keywords, they have fixed meaning in Java. Any Java programmer must know and use them. Most of the remaining English words, while not part of the Java language per se, are defined by a set of software modules that come with Java. They are avoidable in theory, but not in practice. Java programs create and manipulate data entities ("objects") that are assigned to ("instantiate") specific "classes." Each class defines the operations that can be performed on the objects that belong to it. Objects, classes, and operations are all given names ("variable names" or "function names"), which the programmers can use to invoke them. While the programmers can and do define their own classes for their specific situations and can name those however they please, much of the standard functionality is supported by classes that are packaged together with Java. Not surprisingly, such classes and the operations associated with them were named in English, by programmers working for Sun Microsystems (split mostly between California and India).

For example, when the programmer needs to perform date calculations, he can do this by obtaining an object of class "Calendar." Since Java provides a variety of calendars, in this case the programmer chooses a more specific subclass: "GregorianCalendar." To customize the calendar for Brazilian Portuguese, the programmer requests that the new calendar object be created with a Brazilian Portuguese "locale." To do this, he first creates an object of class "Locale" and then refers to this object when making a request for a new calendar.[6] The end result is a command that includes seven English words:

```
Calendar cal = GregorianCalendar.getInstance(new Locale("pt_BR"));
```

To rewrite this somewhat closer to normal English, the line says: "Create a new Locale specification for Brazilian Portuguese ('pt_Br') and then use this Locale specification to set up an instance of a Gregorian Calendar. The resulting object will be a kind of Calendar and shall be named 'cal.'"

The only point at which the programmer could opt for Portuguese is when deciding on a name for the new object that is being created. In this case he used a language-neutral abbreviation "cal," that could stand equally for "calendar" and "calendário." The four names that he introduced in the rest of the code ("weekConvert," "weekday," "start," and "oldDate") are in English. The programmer *could* have chosen Portuguese names for them. However, doing so would turn the code into an odd mixture of two languages, as the programmers often told me. It would also require the programmer to remember whether each particular object's name is English or Portuguese: is it "oldDate" or "dataVelha"?

As we will see in this chapter, however, the choice of language cannot be understood just as a matter of simple practicality. Or, rather, the specific logic of practicality that is invoked in a particular situation depends on more subtle factors. I start my discussion with a few more episodes from Alta. These examples present the use of English and Portuguese in relatively unproblematic situations. I then look at the more complicated case of Lua, a programming language developed in Rio de Janeiro, which until recently had no Portuguese documentation. I follow this story with a discussion of Kepler—a small project based on Lua, which straddles the two worlds in a yet more complicated way. (Alta, Lua, and Kepler are all explored in more detail in subsequent chapters.) I then look at how the developers acquire the knowledge of English, and discuss the social differences that English proficiency marks. Finally, I look at some of the ways in which the developers express resistance to *gringo* dominance.

It's Just More Natural

Looking at Alta's code some time later, while working on a small task assigned to me by "Fabio," one of Alta's managers, I did find examples of Portuguese variable and operation names. A few days after that, as I was watching Fabio draw a diagram of Alta's next application, entering English field names like "price" and "quantity," I decided to ask him about the mix of languages that I had seen in the code. Fabio seemed surprised. *It's supposed to be all in English*, he said. He then explained: *The Portuguese names were just someone's mistake. It all should be in English, except for the database tables.*

"Except for the database tables?" I made a surprised face. *Yes, the Java code should be in English*, Fabio said again. *But the database tables should be in Portuguese.* "Mauricio," sitting at the adjacent table, turned around in his swivel chair. "This is really stupid," he said. Mauricio, in his late twenties, was one of the developers on Fabio's team; he usually stayed quiet, so I knew that this had to be a topic he felt strongly about. *Having class names and database tables use different names makes no sense*, said Mauricio. He explained that there are many software applications that assume that the names of the tables and classes correspond, and a mismatch between the database table names and the Java class names is often a source of endless trouble.

The two started discussing why one could possibly want class names to be in English and database tables in Portuguese. Mauricio's position was that it should all be in English, while Fabio explained that they simply did not have a say about the database tables. Those were administered by "Intermercado," Alta's large client, and had to be in Portuguese. "Why? Because the database guys don't know English? What are they doing there then?" insisted Mauricio. Yes, Fabio explained, the database guys may or may not know English. In either case, this was not for him or Mauricio to decide.

"But why should the *code* be in English?" I asked. "Good question," said Fabio. "I ask this myself sometimes." *But it is just more natural this way*, he explained. *The programming languages themselves are in English.*

Listening to Fabio's comment about programming languages being "in English," I remembered a conversation that I had a week earlier with Rodrigo Miranda, a coordinator of a Rio-based open source project whom

I introduced in chapter 0. In the early 1990s Rodrigo worked in software translation. At some point a call came: Microsoft needed someone in Brazil to translate Excel's Visual Basic into Portuguese. Rodrigo could barely contain his laughter when he was telling me the story. *Yes*, he said, *they literally wanted to translate all the keywords. They wanted to make it "se" instead of "if," for example.* Despite the prospect of making good money quickly, Rodrigo told me that he tried to dissuade Microsoft from doing this. When they decided to go ahead with the project, however, he agreed to do it—the money was too good to pass up. The Portuguese version of Visual Basic failed miserably, much as Rodrigo expected it to. (The money he got for it, though, paid for a new computer.)

I told the story to Fabio and Mauricio. Too young to have witnessed the fiasco, they found it most entertaining, rolling their eyes. *This must be one of the stupidest ideas ever!* they exclaimed at the same time. *How would you even do it?* asked Fabio. *How would you translate "DIM?"* he continued, referring to one of Visual Basic's keywords. *What does DIM stand for anyway? Dimension? So, perhaps it would be "Dimensão." This would be so strange and so verbose!* I responded by pointing out that "Dimensão" could be abbreviated just like "Dimension" was—in fact, to the same "DIM." *I suppose*, conceded Fabio. *But Portuguese just isn't a good language for programming languages. The grammar is too complex. What would you write in the end of the function? "Retorno"? "Retorne"? "Retornar"?* Fabio rattled off several forms of the Portuguese verb "to return." In English it all makes more sense, he concluded.

Fabio's comments about the idea of using Portuguese words as keywords in a programming language do not merely acknowledge the de facto dominance of English in software, but also *naturalize* this dominance. The fact that Alta's programs are written "in English" becomes not a result of a historical contingency—English being the language of the country that emerged as the economic superpower after World War II—but rather a natural state of affairs having to do with the relative complexity of English and Portuguese grammar. We need to note, however, that the explanation simultaneously reserves a role for Portuguese, albeit *outside* the software domain, in situations where its grammatical complexity (and, as my other interviewees note, its *poetic* beauty) could be an asset. In this way, Portuguese is not altogether deprecated—it just has no place in software.

Lua: If or Se?

The question of whether it would make sense to use Portuguese keywords in a programming language had also come up two months before my conversation with Fabio and Mauricio, when I was interviewing Roberto Ierusalimschy, a professor at PUC-Rio, one of Rio's premier universities. In the early 1990s Roberto (following my interviewees, I will refer to him by first name) and two of his colleagues designed a programming language called "Lua," ostensibly for the needs of a specific project being done for a particular large client of Tecgraf, a PUC research laboratory where Roberto worked at the time. By the late 1990s, Lua had become fairly popular outside Brazil for certain kinds of software applications. To understand Lua's success we need to look at how it got "disembedded" (Giddens 1991) from its local context and became "portable" to the unknown contexts of future use abroad. I explore this process of disembedding in much detail in chapter 7. Here, however, I look briefly at a particular aspect of this disembedding: Lua's use of English.

"Lua" means "moon" in Portuguese. "In our language it is a very beautiful word," Roberto wrote on the *lua-l* mailing list in the late 1990s. This name, however, is Lua's only connection with the Portuguese language, and a dubious one at that. "Lua" is also a pun on "SOL"—the name of Lua's predecessor; "SOL" means "sun" in Portuguese, but is at the same time an English abbreviation for "Simple Object Language." (The name is thus a bilingual pun, reflecting the love of wordplay that permeates the software culture.) At the time of our interview, Lua's documentation was available only in English. All of several books written about Lua were written in English. (The most popular one, written by Roberto himself, had been translated into German and Korean but not into Portuguese.) Lua's manual became available in Portuguese only in September 2007, six months after our interview and ten days after a Russian version was released. It perhaps would not surprise the reader when I say that Lua's keywords are all based on English words. Or to put it differently, Lua uses "standard" keywords, such as "function," "if . . . then . . . else," and "return."

When the first version of Lua was being developed in 1993, the choice of language was seriously discussed, if briefly:

Roberto: I remember that we discussed a lot about both error messages and reserved words. There were people, even me, that talked about . . . that maybe instead of "if" we should use "se" and use "enquanto" instead of "while." And we just decided that this is not English—this is reserved words. Someone said that, I don't remember who: those aren't even quite English words, even for English people they couldn't . . . that they were picked by European people who didn't speak English properly. [Laughs.] But anyway, so we decided to stick with the usual reserved words. And I think that the error messages went together; they should be in English, it would be strange to write "while . . ." and then get "Erro na linha . . ." So *maybe* comments were in English for the same reason. I really don't remember. I can maybe try to find some . . . but I think that I usually already wrote many things in English.

Roberto's story brings together a number of reasons for using English, including the technical difficulties with using Portuguese, habit, a desire for consistency, and a justification that English words that appear in code aren't really English. Combined, those factors convinced the authors that using English made more sense.

Lua was, for the most part, originally developed for local use and people who would program in Lua could be assumed to know Portuguese better than English. As the authors of Lua repeatedly assert in publications and interviews, Lua's later success came as a major surprise. At the same time, at the earliest stage of development the team made choices that left open the possibility (however unlikely) of this success, preemptively disembedding Lua from the local language, avoiding a tie that would have forever limited its use to Portuguese-speaking places. I believe that such preemptive disembedding is often driven by what I call "subvocal imagination": imagined futures that are treated as too unlikely to be publicly presented as a rationale for action, but that nonetheless can affect action profoundly. I discuss this notion in more detail in chapters 6 and 8.

The day after our interview, I received a message from Roberto, in which he told me that after our conversation he went looking through old files, finding that while Lua and SOL code were written "in English" from the very beginning, test files for them were in Portuguese until 2003, over a ten-year period. This distinction is easy to understand considering the public nature of Lua's code and the *private* nature of the test files; the practice of using Portuguese for test files and for other "private" code appears to be common among developers who use English for code that is more likely to be seen by others. In some cases this appears to reflect the fact that writing in Portuguese is simply easier, even for those Portuguese speakers who speak English quite fluently. They may therefore choose Portuguese in situations where the resulting "ugliness" of the code is not going to be observed. When writing code for private use, the developers also do not need to worry whether their use of Portuguese might be interpreted as showing a lack of English proficiency. Using Portuguese in private code, however, can also be a way of *marking* the code as private.

It is worth noting that the situation with Lua's test files had to change eventually, as Lua's development made small steps toward the open source development model. This transition demonstrates one of the new

challenges that open source development brings to peripheral participants. Open source development blurs the line between public and private,
because the users of the software often come to expect to have access to the
author's complete working environment rather than just the final product.
Reliance on the local language becomes problematic even for such things
as tests, since the open source logic demands that such code should also be
rendered public. After the users started requesting Roberto's full collection
of test scripts, he eventually released them. While the scripts were originally
released as they were written, with all the Portuguese in them, he translated
them into English for the next release of Lua.

The Lua Book

Much like Lua's code, its original manual of eighteen pages was also written in English, as was the very first paper about the language, presented at
a Brazilian conference in 1993. The conference accepted papers in English,
Portuguese, or Spanish, and the choice of language did not make any difference at the time as to how the paper counted toward the researcher's pro-
ductivity metrics. Nonetheless, there were several good reasons to choose
English. The first one concerned the larger audience that could be reached
by an English publication, or, rather the exceedingly small size of the Portuguese audience:

Roberto: In Brazil maybe there are four or five people who are going to
read what I write. It's not a problem of Portuguese, it's a problem of any
language. You must write in a language that everyone can read, unfortunately, or fortunately, because the number of people is so small. There is
no point of writing a technical paper in Portuguese, or in Spanish, or in
French, or in German, or in whatever language.

For this reason, Roberto nearly always wrote his papers in English. "I usually prefer to write papers in English," he said. Earlier in his career, Roberto
explained, he used Portuguese for the relatively unimportant papers ("when
it was very fast"). Now he prefers English in all cases.

Roberto then mentioned a different factor, the technical challenges of
writing papers in Portuguese. Like most computer scientists in the United
States, Roberto uses a software product called LaTeX to write his papers.
LaTeX is essentially a programming language designed for typesetting
documents and, like most programming languages, it cannot easily handle
non-English letters or accents. "There are a lot of packages to solve that but
I do not have them installed properly," Roberto explained. "Or sometimes

they *are* installed properly but I change the version and I only discover two months, three months later that they are not working. Apart from emails I almost never write anything in Portuguese." English is thus intertwined with the practice of academic computer science writing in a very material way, being assumed by many of the tools on which the practice relies, not unlike the way it is intertwined with the practice of software development itself.

Nearly all publications about Lua were thus written in English. Around 1996 one such article, in a popular American computer programming magazine, attracted a substantial number of questions, which led Roberto and his collaborators to start a mailing list, *lua-l*. (No mailing list was necessary before that since all users worked in the same place.) From the beginning, most of the subscribers were foreign and the discussion was conducted in English, though messages sent in other languages occasionally popped up and were typically met with friendly amusement or curiosity rather than disapproval. In 2002, when the community had grown substantially, a question was raised whether a separate Portuguese list was needed or whether the list should be declared officially "multilingual." In response to this discussion Roberto pointed out that Brazilians comprised only around 15 percent of the list. "Our second 'minority' are 10% of German speakers," he noted in parentheses. "Whether we like it or not, the only language we can all communicate is English," he concluded.

In 1999 the mailing list was informed that Roberto was working on a book about Lua. "Do us poor language-handicapped folks in the States have a chance of being able to read it?" asked an American list member. "I hope so," responded Roberto. "I am writing it as close to English as I can:-)" The book was in fact written in English. When *Programming in Lua* was finally ready in 2003, Roberto asked the list for ideas on how to publish it. The list members responded with suggestions and offers to proofread the book or to represent Roberto in the United States. (It was largely taken for granted that the book would be published there.) Eventually, Roberto self-published the book via a print-on-demand service that also acted as a distributor and could thus satisfy the number one requirement: that the book become available for purchase on Amazon.com.

Roberto's announcement of the book's availability on Amazon came with a note in parentheses, saying that Roberto was trying to get a smaller batch printed in Brazil to make the book available at a lower price. This plan, however, never materialized. An attempt to get the book into some of the Brazilian stores was also unsuccessful. The stores did not want to buy the book from a foreign distributor, and insisted on ordering it directly from the "publisher" instead. Under the print-on-demand setup, this would mean buying the books directly from Roberto, who would have to set up a company to do proper accounting of sales and expenses. This ordeal of starting a company is not taken lightly in Brazil: it is hard to start one, expensive to maintain it, and nearly impossible to close. Foreign publishers avoid this problem by maintaining their tax home abroad. Roberto could avoid it by staying out of the Brazilian market and letting the potential Brazilian readers just order the book from the United States, something many of them are quite used to. In many ways, self-publishing the book in the United States and making it available on Amazon.com might after all be the most efficient way to reach *Brazilian* readers.

While one could not buy a copy of *Programming in Lua* in any of Rio's stores (not even in PUC's own bookstore), I did manage to obtain my copy in Brazil without resorting to Amazon.com. I simply asked for one at the end of the interview. The book came from a small stack in Roberto's office, and I paid for it in cash. This local transaction exemplified a broader pattern. While Lua's success was dependent on both local and global ties, its local ties mostly were (and still are) truly local, often limited to the small network of Roberto's students and colleagues, with relatively few connections at the national or even city level.

By the spring of 2007 the book had been translated into German and Korean (and later also into Mandarin), but no Portuguese translation was available (as is still the case today). Rodrigo Miranda, Roberto's former PhD student and a tireless promoter of Lua, was trying unsuccessfully to organize a translation. A half-completed translation by his friend "Renato" died with Renato's hard drive. Rodrigo had not yet had the courage to tell Roberto about the loss. Roberto was hardly counting on this effort to come to fruition, however, and had few hopes for successful publication of the translation if it were ever finished. He would love to have the book translated, he explained, but he was not going to translate it himself and had little interest in begging others to do the work. After all, he did not have to do this to get the book published in German or Korean.

Kepler's Wiki

Despite its notable popularity, Lua is a niche language, which has gained substantial popularity in a particular set of applications, typically those requiring performance, simplicity, and close interaction with software written in C. In California, Lua is primarily used in the development of computer games or software for small devices. Both of those applications are too specialized for Rio's small market, which depends primarily on building web applications for local businesses. Rodrigo Miranda, a former student of Roberto, dedicated a decade of his life to the uphill battle of turning Lua into a platform for developing web applications. In 2007, I spent four months following Rodrigo's project (named Kepler) from inside his office in Copacabana—a story that I tell in more detail in chapter 8.

In early March, after a week in Rodrigo's office, I was looking for more hands-on involvement with the project. (The project was so highly "virtual" that just observing what was happening in the "real world" was rather uneventful.) I decided to start by helping with the project's web site. Over the next two weeks, Rodrigo and I had some long discussions about what needed to be done. One of the things that Rodrigo asked for was to have the web site use a wiki, allowing visitors to edit the content.

A couple of weeks after our first discussion about the web site, I realized that having designed most of it, we had not thought at all about where exactly the Portuguese version of the documentation would go. The existing Kepler web site did have Portuguese documentation, though it was spotty and the reader was often sent back to the English version. I did remember, though, that Rodrigo had talked about the need for Portuguese documentation and even that a software developer I had not yet met—"Rodolfo"—was working on it. Running the web site as a wiki was going to create additional challenges: we would need to figure out how to keep the English and Portuguese content synchronized, even as the users would be potentially editing each version separately. I brought this issue up while talking to Rodrigo over instant messenger. His response confused me. Rodrigo seemed surprisingly disinterested in the Portuguese documentation, even as he reaffirmed that Rodolfo was working on it. There was no need to worry about whether the two versions would remain synchronized, he told me, though perhaps Rodolfo or "the community" could take care of it.

It was only a few months later that I managed to fully make sense of this conversation. Kepler was supported by FINEP, a Brazilian agency that funds research projects in industry, and by "Nas Nuvens," a company that belonged to Rodrigo's brother "João," which was cosponsoring Kepler's development. Nas Nuvens hoped to eventually offer Kepler-based solutions to its local clients and needed documentation in Portuguese. For this reason, it had requested FINEP funds for writing Portuguese documentation as one of Kepler's subprojects. There also was an extra benefit: subcontracting documentation work to Rodolfo created an opportunity to build a relationship with Rodolfo's organization, a research institute near Rio de Janeiro. As I learned later, Rodolfo had recently sent Rodrigo a draft, but Rodrigo had not found time to read it. In April 2007, Portuguese documentation had not yet become a high priority: Kepler had to be finished first, and for that to happen it needed the attention of high-end developers who could contribute to the project. This meant documentation in English.

A week after our discussion over instant messenger, we found ourselves rethinking our initial choice of the wiki software. During that deliberation, I decided to try writing a simple wiki engine in Kepler—after all, Kepler was a platform for web applications. A few days later, with the initial version of the software in hand, we revisited the question of Portuguese pages, talking about ways of interlinking the Portuguese pages with their English equivalents.

The URLs of wiki pages are usually based on their titles, which makes it easier to create links to them. The URL of a page called "Introduction" would end in "/Introduction." If the same were true for Portuguese pages, then the corresponding Portuguese page would show up under "/Introdução." This created a minor challenge for figuring out how to cross-link the corresponding pages in the two languages. I asked Rodrigo what he thought. Rodrigo seemed shocked at the very idea of using Portuguese in URLs. He wanted simple correspondence between the pages, and he did not want any accents in URLs. And he would not want pages with names like "Introducao." Rodrigo carefully enunciated the hard "c" and the "a" of "Introducao," to show the effect of the missing accents by making the end of the word sound like English "cow" rather than like "sung" for "-ção." The page, he continued, should be called "Introduction." Calling it "Introdução" or "Introducao" would be just as ridiculous as using Portuguese words for Lua or Kepler keywords. I noted to myself that using Portuguese keywords in a programming language apparently set a benchmark for "ridiculous" for Rodrigo, but kept the thought to myself, instead turning the conversation to the remaining practical issue—how to insert links to English page names in the text of Portuguese documentation.

Another week later, Rodrigo and I were about to leave the office when we got chased by João, who wanted to know about the state of the Portuguese documentation. Rodrigo replied that he had glanced at Rodolfo's document and that it was good enough for now. (The FINEP project was not due for another nine months.) "You are now talking like Roberto," said João in frustration. Rodrigo got defensive. *It's not quite the same*, he said. "I am not saying that there *shouldn't* be Portuguese documentation," he explained. There were just better things to worry about. The elevator arrived, giving us a chance to escape and leave João behind to worry about the Portuguese documentation alone.

Kepler's complicated relationship with Portuguese documentation reflected the complex way in which Kepler connected to two different worlds: first, the local world of Brazilian clients and partners, and, second, the foreign world of software development in which Kepler was trying to find a place. Roberto's situation was simpler. As a Brazilian academic researcher, Roberto was shielded from the local business world. For him, Portuguese documentation was a luxury, "a nice-to-have feature," to use a programming term. Rodrigo's approach, on the other hand, puzzled me for many months, as Rodrigo went back and forth between spending time to figure out exactly how the Portuguese documentation would work and seemingly questioning any need for it. As I came to see it eventually, for Rodrigo, the issue of Portuguese documentation was a balancing act between conflicting commitments. Kepler's destiny was tied to that of Nas Nuvens. As we will see shortly, some of Nas Nuvens's programmers could read English only with difficulty. Additionally, like many other software companies in Rio, Nas Nuvens also was dependent on local clients and had to build other local alliances. Yet the small funds that the clients and FINEP could provide did not allow Rodrigo to obtain what he needed most for his project: expertise. He looked for such expertise in two places: abroad, by trying to recruit the invisible programmers from the global Lua community, and among people at PUC, many of whom considered it silly to spend time on Portuguese documentation, not to mention worrying about accents in page names.

Learning English

The fact that Roberto Ierusalimschy's *Programming in Lua* was available only in English did inconvenience some of the developers working with the language. For "Luciano," one of the programmers at Nas Nuvens, *Programming in Lua* was the first and the only (as of 2007) book he had read in English. The task took him a long time, he told me, and he was reluctant to try it with another full-length English book. "It takes forever," he said.

Luciano was twenty years old at the time of our interview and came from a lower-middle-class background.[7] He went to a public school, where he had English classes, which he said had taught him nothing. ("English for twelve-year- olds. Doesn't count.") What *did* teach him English, he said, were role-playing games (RPGs) on a computer:

Luciano: Games, RPG games, you know? I played a lot—the RPG games. The games gave me the minimum and then . . . For example, to buy things, you need to apply [. . .] the word "buy," and the word "sell," you know? [. . .]

If you need to open this door, it's "open the door." You apply: "open the door." [Pause.] It's . . . it's *buy* [says in English], right?

Luciano was neither the first nor the last person to credit computer games for an introduction to English.

Since computer games serve as a common entry point into the tech world for many Brazilian men, Luciano's learning of English was from the very beginning fueled by his growing involvement with tech culture. Around the same time he started playing computer RPGs, Luciano began fixing computers as a part-time job, then got interested in the Internet, got himself a computer, and started learning HTML. He got a job in tech support, and started learning Linux and PHP—the latter a popular programming language for web applications. His source of learning was translated books and online forums in Portuguese ("some forum mentioned PHP," he says). It did not take him long to discover, however, that most of the material he needed was in English, and he started trying to make sense of the English materials on the web.

At the time of our interview, Luciano frequently used English materials, but remained selective, considering the difficulty of using such documents against the expected value of the information contained in them. When looking for an answer to a specific question, English would often be his choice. When wanting to learn a topic in more depth by engaging with a longer text, however, he would often seek something in Portuguese. "With English it is really easy for me to get lost," he explained. "One word will totally change the meaning of everything. I can't be sure I got it right."

The need to solve practical problems and an interest in engaging with the tech culture are not the only motivations for learning English. The language is also widely seen as more generally providing access to the larger world. On many occasions, for example, I heard the developers say that they learned English because they wanted to understand the lyrics of rock or heavy metal music. (A handful in fact claimed to have learned the language primarily through lyrics.) Others talked about American movies or non-computer-based role-playing games. (In the case of the latter, they would of course speak Portuguese to fellow players, but would often have to rely on English manuals for the game.)

This connection between English, games, software, and the larger "global" culture illustrates the parallel replication of practice noted in the previous chapter: re-creation of the practice of software in Brazil is in this case aided by the fact that some of the skills that this practice requires, such as English proficiency, are also employed in *other* activities. This, of course, has much to do with the fact that software documentation, games, and a lot of the music that Brazilians listen to tend to originate from roughly the same part of the world.

While Luciano had picked up a good amount of English by the time we met in 2007, his English skills were lower than those of many other developers I met in Rio, who could often not only read English but also speak it with relative fluency. While a few of those developers claimed to have learned English on their own, and a few others credited exceptionally good private schools, the most common pathway to English proficiency involved private English courses. For well-off Rio families, enrolling their kids in such courses is an obvious choice, and such kids typically start their course well before college. (Some of my interviewees started private English courses as early as age ten.) Those coming from families with fewer resources have to wait until they can afford the course, in terms of both money for tuition and the time taken off from work or undergraduate studies.

One of the other developers at Nas Nuvens told me:

Pedro: Because what I did is I didn't do my undergraduate program in four years. I did it in four and a half years. I did it in four and a half because I decided to reduce the workload to study English. So, I reduced the load to study English, to study English for a year and a half, with a commercial English course. *If you want, we can talk.* [Says in English. Laughs.] My English is still, still . . . I understand better than I speak. [. . .] Because during the sixth semester I started thinking: if I don't study English I won't be able to do a master's degree. [. . .] Because with the master's program here—they [expect] that a person would be fluent . . . not fluent, but would be capable of reading. [. . .] Must know to read and write English.

As Pedro noted, while Brazilian universities rarely teach English, the better ones frequently expect the students to be able to read it, and this is especially true for graduate programs—even more so in computer science. An investment in English thus offers not just an opportunity for drawing on remote resources but also access to *local* educational resources. In Pedro's case, the investment seemed to pay off: since our conversation in 2007, he successfully entered into a master's program at PUC, completed it, and is currently contemplating a PhD abroad.

The flip side of opportunities offered by English is the stigma that is often associated with the lack of English proficiency. To a large extent, this may be a matter of simple realism: not being able to read technical documents in English does make it more difficult to stay abreast of computer knowledge. At the same time, English proficiency also functions as a marker for class distinctions, since it requires either access to financial resources, cultural capital available to the children growing up in upper-class families, or both. Mentioning that a particular developer "cannot even read English" consequently becomes one of the worst assessments, one that connotes a lot more than a simple lack of English proficiency.

It is perhaps not surprising that I often found my interviewees quite reluctant to admit their use of Portuguese. Portuguese books that I saw on developers' desks inevitably turned out to be bought for the sake of being lent to friends or colleagues. To the extent that developers were willing to discuss their difficulties with English, they often appealed to my understanding as a fellow nonnative speaker of English. "Do you really want to know?" one of my interviewees, "Edmundo," asked me after I inquired in what cases he used Portuguese in Google queries. (Edmundo spoke reasonably good English and we even conducted a part of our interview in it before switching to Portuguese.) "Do you really want to know?" he asked again. He then finally said: "When I am too tired to write in English, then I enter it in Portuguese." Edmundo chuckled. "Because it's not my native language," he then explained. "It's not my native language. It's like you." He asked me how many years I had lived in the United States. I told him it had been around ten years. "Ten years, damn!" said Edmundo. "And still reading in Russian is much easier for you than reading in English."

The Speakers and the Nonspeakers

A few months before my interview with Luciano, I was sitting in the office of Nas Nuvens, in the room that I shared with Rodrigo Miranda, when Luciano knocked on the door. He stepped inside and said something to Rodrigo. Rodrigo responded by pointing to me as the person to ask. As it turned out, the question concerned Linux and I seemed to be the resident Linux expert. After a brief discussion, I concluded that the question was out of my competence and suggested that Luciano ask "Alan," an active Kepler developer who used to be at PUC but recently moved from Rio to Porto Alegre. I remembered, however, that a few weeks earlier Rodrigo had told me that he wanted to start running Kepler as a "real" open source project, and that this would involve routing more communication through the mailing list and relying less on face-to-face interactions or private email. So I told Luciano that Rodrigo would probably prefer if he asked this question on the Kepler list rather than emailing Alan directly.

Rodrigo nodded. Luciano looked at him in a bit of disbelief. *You are not going to make me do that, right?* said his face. *Yes*, Rodrigo responded to Luciano's silent question. *Write to the list.* I did not catch Luciano's response, but the prospect of writing to a public mailing list in *English* did not seem to appeal to him.[8] I was about to volunteer to help Luciano compose the email when Rodrigo sighed and said: "Write it, email it to me, I will translate it for you." Luciano nodded and left. A bit later, a message from Luciano arrived via the mailing list.[9]

A few weeks after the incident with Luciano's email, I was at Nas Nuvens's office again, slouching in a beanbag (or *puff*, as it is called in Brazil), while Rodrigo stood in front of me. I was trying to explain a problem I thought we had with a code example we were working on together. I was speaking Portuguese and I stumbled as I searched for the Portuguese equivalent of "smart quotes." Rodrigo was following my line of thought, however, and completed my sentence for me: "smart quotes." He used the English phrase, but pronounced it as if in Carioca Portuguese: "ishmahchi quotish." After we concluded this discussion, I switched to English and asked Rodrigo why he would say "ishmahchi quotish," if he knew how to pronounce this phrase in English.

Rodrigo sat down on the *puff* next to me, leaned back, and took a deep breath. *Many people don't know enough English to know how to say it right*, he said after a pause. I understand this, I responded, but *you* know how to say it. Why do you say it this way? *My English is not so good, actually*, he said. "This is ridiculous," I thought to myself. Rodrigo's English was almost as strong as my own. (Neither of us was a native English speaker, after all.) *Listen*, I said, *maybe your English is not perfect, but you know how to say "smart quotes."* It is clear, I continued, that people who are fluent in English often say English words with a strong Portuguese accent when using them in a Portuguese sentence.

Okay, said Rodrigo, *I'll tell you why. It's because Luciano was in the room. I tend to speak this way when there are "nonspeakers" in the room*, he explained. There is a thing about using English in a "politically correct" way, he continued. *When you use English, you don't want to make it sound like you think you are better than other people and if you speak overly correct English people might think that. If I say "ishmahchi quotish," it makes me just one of the guys. It's a way of "making fun of English, making it less elitist,"* he concluded. Somewhat in disbelief, I asked Rodrigo if he had just come up with this theory on the spot. *No*, he replied, *I first thought about this twenty years ago, in high school.*

A linguist would disagree with Rodrigo's explanation, pointing out that saying "ishmahchi quotish" can be explained as simply a matter of unconsciously adjusting the pronunciation of an English phrase to the phonological context of the Portuguese sentence into which it was inserted (see chapter 6 in Grosjean 1982). What is notable, however, is that the question I asked was something Rodrigo has thought about. The divide created by the differences in English proficiency cuts both ways. In some cases, a proficient English speaker who spends much of his time interacting with those who speak it less may want to downplay his English skills, to simply be "one of the guys."

The World Language or the Gringo Language?

I was at Nas Nuvens when an email arrived from Rodrigo, sent to me, his friend Renato, and the mailing list that includes all of Nas Nuvens's programmers. It was a link to a blog post entitled "No Mundo da Lua." The Portuguese title was a pun—it could be translated as "In the World of Lua" or "With Heads in the Clouds" (literally, "in the world of the Moon"). The blog post lamented the fact that Lua was unknown in Brazil and that *Programming in Lua* was available in English, German, and Korean, but not in Portuguese. Ironically, it then directed readers to an article about Lua in the *English* Wikipedia.

The incident left me curious about the relative length of Wikipedia articles about Lua in different languages. I spent some time looking at them, compiling a table.[10] I was not particularly surprised to see Portuguese below Korean and Spanish. Rodrigo walked into the office just when I was getting the word count for the Esperanto version. I asked him to make a bet: would the Esperanto article on Lua be longer or shorter than the Portuguese one (visually they appeared quite similar). Rodrigo bet on Portuguese, without too much excitement. The Portuguese article did turn out to be longer, though just barely. I announced the result to Rodrigo. He looked at the Esperanto article in disbelief and seemingly a bit irritated. *I just don't get it*, he said.

As it turned out, Rodrigo was not irked by the fact that Portuguese nearly "lost" to an invented language, but by the fact that people waste their time on Esperanto. "Anyone who can speak Esperanto can also speak English," he explained. *If they can speak English, why do they bother with Esperanto?* I tried to summarize my understanding of the motivation behind Esperanto: some people believe in having a neutral language to communicate in. We continued the conversation as we headed out to get some food. I tried a few other arguments that I could remember from a book on Esperanto that I had read back in the Soviet Union. Rodrigo was not fully convinced.

I speak English because it is practical, explained Rodrigo. *A while back I figured out that I could only learn one foreign language. English was the best option since that was the language spoken by the most people.* I asked him why he did not learn Chinese, if the number of speakers was the determining factor. True, Rodrigo agreed, but Miami was closer than China. He added something about McDonald's. There you go, I said, it was not about the number of speakers, but about McDonald's, Miami, and Disney World. Or, to put it differently, it was about cultural dominance. If Buddhist temples were more important to you at the time than McDonald's, I concluded, maybe you would have tried to study Chinese after all, alluding to Rodrigo's Mandarin-speaking Buddhist friend.

Sure, agreed Rodrigo. *But English is not the same as the United States.* In some years, he continued, the United States will no longer play a dominant role on the world stage, but English will still be the main language. It has nothing to do with the United States or with its culture. "The United States of Canada" will be mostly Spanish-speaking anyway, and nobody is going to care what people speak in "Jesusland," he said, referring to a 2004 US election joke I had told him earlier, which envisioned an alternative division of North America. English will thus no longer be seen as the language of the United States, just a means of international communication.

Rodrigo was speaking in a somewhat humorous tone, and his words seemed to be carefully chosen to express neither hope nor disappointment at the eventual demise of the United States that he was foretelling. He talked as an indifferent, if curious and somewhat amused, observer. This was far from the only time when the future demise of the United States came up as a conversation topic during my time in Brazil, however.

About a month later, I was having lunch with a group of Alta programmers. Having talked about Linux music players and recent gadgets on the way to the restaurant, we made our way through a number of nerdy topics, arriving eventually at the issue of measuring temperature in Fahrenheit degrees. *How stupid is that?* said one programmer. Others nodded in agreement. *The whole world uses the metric system, except for the United States*, he continued. *Why can't they act like a normal country?* Others nodded again. *And then we end up using the stupid American measures too*, interjected another developer. *Like measuring monitor sizes in inches!* (Brazilians measure TV and monitor sizes in "polegadas," a term that otherwise comes up only in translated books.) *But one day the United States will decay, and perhaps the idiotic measurement system will facilitate this.*

"I hope," said another developer, Marcos, "to live long enough to see three institutions go down. The first one is the United States. The second one is Rede Globo, which won't take that long. The third one is Microsoft." Marcos then moved on to stronger imagery, talking about how each of those needs to be "destroyed." (The picture of destruction was painted most vividly for Rede Globo—the country's main news network that middle-class Brazilians love to hate. The disdain they expressed for the United

States seemed bleak in comparison.) Other developers made supporting comments regarding all three. Nobody seemed to question that this was about the right list. The conversation did not linger on this topic for very long, however. After a few clichéd (and seemingly pro forma) curses toward Microsoft, the discussion moved to the activities of the Gates Foundation, then quickly to finance, and the rest of the lunch was spent talking about personal investment in stocks.

Open expressions of hope for the "destruction" of the United States are not something a US-affiliated researcher hears every day, of course. What I did hear regularly, however, were the more mild references to *gringos*, typically accompanied with the lightest touch of resentment, and immediately retracted upon any interrogation. From my first days in Brazil, I was continuously surprised by the extent to which the people I interacted with closely (software developers or not) highlighted my *Russian* origin when presenting me to others, bringing up my connection to the United States only when wanting a joke at my expense. (The situation was a bit different in the more formal interactions, where being "from the United States" was more valued, though even there "from Berkeley" or "from California" seemed to be preferred.) While part of the preference for seeing me as being "from Russia" rather than "from the United States" no doubt had to do with the curiosity toward a distant and somewhat mythical country, it seemed clear to me that such identification was also meant to allow me to not be seen as "a gringo."

The resentment toward gringos, however, is never expressed as resentment toward *English* among the software professionals that I interviewed. Or, to be more precise, it is never expressed as resentment toward English *in the context of software*. Lamenting the inappropriate love of English among Brazilians broadly is more common. In one of our conversations about Lua, for example, Roberto talked about Brazilians' preference for foreign things in general. "People here love to use English phrases," he said. "In the beginning of the nineteenth century everyone loved to speak French. In the beginning of the twentieth century it was English but from Britain. Now it's English from America." This love for all things foreign, however, should not be confused with *their* use of English, point out Brazilians working with software. The latter is an entirely practical affair. And in the near future their use of English might have nothing to do with the United States anyway.

The situation with language here reflects a broader pattern. As I show in subsequent chapters, being located at the periphery often leads to a lack of cohesion. Peripheral actors all too often feel "stuck" in the wrong place, with the wrong people. Managers say they can get nothing done in Brazil because of the incompetence of the employees. Programmers say you cannot get anywhere because of the clueless managers. Both blame the government and the clients. The clients and the government find their reasons to be dissatisfied with both the workers and the managers of the IT firms. It is in this context that the actors then accuse others of an irrational preference for foreign things, while stressing that their own efforts to establish direct links to the outside are simply a matter of realism.

We must remember, of course, that the distinction between the "pragmatic" use of English and its symbolic use is sometimes real. Highly educated individuals like Roberto and Rodrigo (as well as many of my other interviewees) are in fact sufficiently worldly and global (at least in comparison with many people around them) that they do not usually need to show their worldliness to others—or, rather, they can do this in subtle ways. Their command of English is sufficiently obvious that they do not need to use it in front of others just to make a point (though this always remains an option). In fact, as we saw earlier, they may be more concerned with how to fit in to the world of nonspeakers.

Use of English can thus sometimes be a pragmatic choice, a matter of reaching the largest audience, and at other times a way of establishing social status and flaunting connections with the larger world. It can also mark local connections (between the members of the educated elite or between engineers sharing the same jargon), or draw a distinction between those with and those without education. Use of Portuguese can similarly mark connections or boundaries. As we will see later, much the same can be said about many other types of cultural codes.

* * *

In this chapter we looked at the place of English in the practice of Rio's software professionals, visiting several of the contexts that are explored in more detail in some of the later chapters. We saw that English is accepted by the developers as the professional language of software, even as it remains unambiguously a foreign language in Brazil. English provides developers with an opportunity to create global links, allowing them to draw on remote resources. Those resources include not only written materials, but also the foreign software on which the developers build their work. Such software is intertwined with English to such an extent that the idea of programming "in Portuguese" becomes nearly unimaginable for many of my interviewees. For some, English also provides an opportunity to actively reach out and engage with foreign members of the practice and to recruit them for the developers' own projects.

As much as it can link, English can also divide. Because English proficiency varies among software professionals, being strongly associated with socioeconomic class, heavy reliance on English can introduce language barriers within the local community. Such barriers can be symbolic, marking proficient English speakers as standing apart from the nonspeakers. They can also be simply practical, depriving those who lack English proficiency of products of the English-mediated work by others in their local community. Such barriers affect most obviously those who are left out due to their lack of English skills, but they can also become a problem for the English speakers who face a reduced pool of potential collaborators. Ultimately, those boundaries are often bridged by passing through the centers of the software world: Brazilians wanting to learn Lua would need to learn English.

The daily use of English, generally taken as unproblematic, takes place in the context of a somewhat more complex system of attitudes toward the United States—a country that is often admired but is also criticized and even scorned. The developers often resolve the resulting tensions by discussing English as the professional language of the *global* software community, pointing to its use beyond the United States. They stress that they use English not just for communicating with people in the United States and Britain, but rather to engage with software developers around the world— speakers of German, Finnish, and Polish. They occasionally mock those Brazilians who use English out of love for all things American. Their own use of English, they stress, is a matter of global pragmatism.

The interactions that surround the use of English illustrate a number of points that apply more broadly to peripheral participation in a global practice. Lua's use of English, for example, illustrates the broader notion of disembedding—Lua's separation from its local context would later enable it to travel globally. We saw the parallel replication of practices, with the developers' learning of English facilitated by the fact that English does not just serve as the language of software, but rather provides access to many other English-mediated practices. We saw that Brazilian users of Lua would likely need to approach it through English—but they will also likely need to have *Programming in Lua* physically shipped to them from the United States.

While investigating the use of English by my Brazilian interviewees, we looked at how they come to learn the language. The next chapter looks more broadly at the process of becoming a software developer. After that I take a look at the larger world of software entered by the developers. (I postpone this presentation in recognition of the fact that the young "nerds" we will meet in the next chapter do not always know what kind of world they are entering.) I then return to Alta, Lua, and Kepler in later chapters, looking at each as a different potential configuration of local and global commitments.

3 Nerds from the Baixada and Other Places

"Since I was quite a *nerd*, I spent most of my time in the computer lab," said Mauricio, talking about his high school years while answering my question about how he became a programmer. The word that Mauricio used to describe himself was a borrowing from English, just like many of the other Portuguese words related to software. When written, this word is spelled in Portuguese just like in English: "n-e-r-d." When used in speech in Rio, however, its pronunciation is normally adapted to the phonetics of Carioca Portuguese, resulting in a sequence of sounds that would likely be unrecognizable to most English speakers: "NEH-jee," with a somewhat harder "H" than in English. It has roughly the same meaning as its English cognate, though with a heavier connotation of computer use and often a more derogatory feel.

When I asked Mauricio to explain what he meant by "being a nerd," he seemed puzzled by my question and replied with another English word: "geek," this time pronouncing it just as in English. He liked computers a lot, he explained. He then added: "I wasn't a very social person. I spent more time installing programs than doing other things." For Mauricio and for many of my interviewees, "nerd" is a basic concept and my questions about its meaning were quite often met with a degree of disbelief. They must have been particularly puzzling coming from an interviewer who knew how to program and gave many signs of being a nerd himself. Surely I would know that nerds are people who are not very social and spend a lot of time with computers.

As suggested by Mauricio's example, this simple term often appeared to carry in it a seemingly simple answer to the question of how one becomes a software developer. For many of my interviewees, software work is simply a natural career choice for a nerd. But how does one become a nerd then? For many developers, this seemed to be silly question too. One does not *become* a nerd. It is just something you *are*, something you discover about yourself in childhood. Some developers argued that nerds are actually born with different brains, perhaps with a mild form of Asperger's syndrome. Looking at developers' stories more closely, however, reveals that becoming a nerd is best understood as a process of a gradually deepening engagement with a world of practice. I explore this process and the eventual transition from being a childhood nerd to a software professional in this chapter.

Even though being "not very social" is a key part of many developers' definitions of being a nerd, the process of becoming a nerd (and later a developer) cannot be understood without considering the individual's engagement with other people. Talking about his nerdy high school years, Mauricio told me the following story:

Mauricio: He [the teacher] would come, give a class, and let people go and the class would go to play soccer. The whole class would leave and we would stay there in the lab. The thing is that *Doom* came out, so . . . The big thing to do was to get a mouse and break it to make a modem cable. To play *Doom* against . . . [each other]. [. . .] The mouse had the right connector—serial. [. . .] It was cheaper to get a mouse, break it and make a cable. It got to a point that we had so much practice with this . . . We would pull it out of the mouse [picks up an imaginary mouse, rips off its cord and removes the imaginary insulation with his teeth], connect the wires, attach . . . It took less than five minutes to make a cable.

Mauricio presented the story as an illustration of the idea of not being "social." (He later explicitly contrasted this to the "social" pastimes of his peers: "playing football, going to the beach, dating.") Yet, he repeatedly talked of "we." Mauricio's learning how to convert a computer mouse into a do-it-yourself serial cable may seem like an example of "not being social" only if we ignore the fact that he and his friends practiced this skill in order to connect their computers and play *together* rather than individually.

Growing up as a nerd is not the only way to become a software developer and I explore some alternative pathways at the end of the chapter. This particular path, however, is not only common but is also important because this is how one is *supposed* to arrive at a software career. Membership in a world of practice often implies acceptance of a collective explanation of why the members engage in it and what makes them choose this particular practice over alternatives. Such explanations may vary between the different worlds: for example, the practice can be understood by people who engage in it as a way of making a living without sacrificing freedom of thought (Willis 1981 on manual workers), as service done for the benefit of other people (Orr 1996 on Xerox technicians), or as disciplined and honest work (Lamont 2000 on white working-class men). In the case of software development, however, the normative answer usually stresses a passion for software born out of childhood fascination with computers. Those who lack such passion are usually careful not to advertise this fact. The role of passion, therefore, must be understood simultaneously as a matter of reality for many developers and as a matter of *mythos* of the community as a whole. It becomes important to keep in mind the stratifying effects of this mythos, as it celebrates the experiences of some members over those of others, often reinforcing boundaries of class, gender, and geography.[1]

Child nerds who take their early steps toward a software career by playing with computers rarely know where exactly this road will take them. One of my interviewees, "Célio," working as a systems analyst for a Rio office of a foreign company at the time of our interview, recalled developing an interest in computing when he got an Atari video game at age six. "So I decided I wanted to do that for my life," explained Célio. "Though I didn't know what 'that' was." The understanding of the nature of "that" which they are joining comes only later and gradually. One of the aspects of this "that" that becomes fully apparent later in their lives is the economic and geographic structure of the world of software.

To facilitate the presentation of the developers' own unfolding understanding of this structure, I do not present in this chapter my own take on it, reserving this discussion for the chapter that follows. Even so, however, it is hard to miss the way in which the developers' entry into the world of software is affected by their position on the world map. We will see the young nerds entering the world of software development *in* a particular place and *from* a particular place. They enter the world of software *in* Rio de Janeiro, in the sense that most of them will practice software in this city for most of their lives. Their experiences of the software world will often be experiences of the software world in Rio de Janeiro. They also, however, enter the world of software *from* Rio de Janeiro (and its suburbs), as they start to understand early on that the local world of software is but a minor site in the larger, *global* world. For this reason, future developers must find ways to transcend—to the extent that is possible—the limits of the local place, becoming members not just of the local software community, but of this larger world as well.

Hanging Around, Mapping Interrupts

"Zé Luís," who also goes by an English nickname "Jason," was in his early thirties when I met him in 2007. Like the majority of my interviewees, he

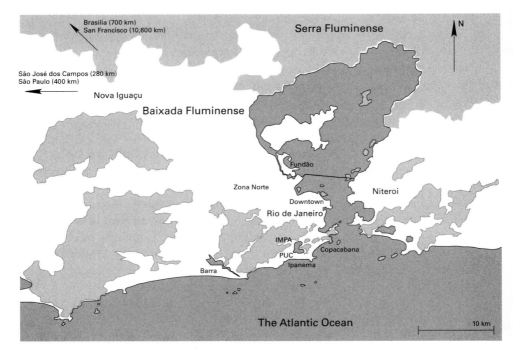

Figure 3.1
Rio de Janeiro and surroundings.

had lived his whole life around Rio de Janeiro. Like many other software developers with a lower-middle-class background, Zé Luís grew up on the outskirts of Rio, in the area called known as "Baixada Fluminense." In his case it was Nova Iguaçu, one of the larger municipalities in the Baixada, forty kilometers northwest of Rio. (Figure 3.1 shows the relative location of the different part of the Rio metropolitan area.) Zé Luís described Nova Iguaçu as "a peripheral city, in a third world country"—a description that would probably also fit other municipalities in the Baixada.

Like many other software developers his age or younger, Jason started his software biography with his childhood:

Yuri: And how did you begin working or doing things with computing?
Jason: I've been doing things with computing since I was eight, eight years old. I started working with small computers using Sinclair logic, which in Brazil were commercialized by the name "TK85." Those were really small computers and my dad bought one of them for me, and I developed little

games on it, and my cousins, who were the same age as I, played those games, suggested changes, and I would go ahead and implement them. I learned BASIC using the manual of the computer, which came with native support for BASIC. So I learned it there more or less by myself, and got really interested. But I didn't pursue this much further. Actually, I wanted to be a writer, to write fiction. I always had rather diverse interests, in different areas. So, it was only years later, when . . . In the eighties, the education system had a series of problems with the government at that time, for a few years. So there were many strikes and they created gaps of sometimes up to four months during the academic year. [. . .] During one of those my dad thought it was important to put me in some sort of course so that I wouldn't lose a year without studying. So he put me in a computer [*informática*] course. [. . .] There in this computer course I was introduced to other technologies: databases, those things. And then eventually got interested in this as a career.

Jason presented his involvement with computers as happening in two phases. At age eight, his dad got him a computer on which Jason learned to program in BASIC. He did not, however, pursue this interest further at the time, returning to programming only much later, when he was fourteen.

This two-step story is remarkably common, and I believe it reflects the developers' desire to establish the time of their *earliest* experience with computers, since engaging with software in early childhood is one of the ways of demonstrating one's credentials in a practice that expects passion and a degree of inborn proclivity. As I learned, I was not the only person asking developers how they got into programming. At least one of my interviewees talked about asking this specific question of all job applicants. He was looking to find people who did not just do programming for a living, he explained to me, but rather those who *loved* to program. Asking them about their entry into the world of software was one way to gauge passion. What he hoped to hear were answers like Jason's. (Others sometimes said they did not need to ask developers such questions, as they could just "see it in their eyes.")

Jason's story of his engagement with computers started with programming computer games for his cousins. Other developers' stories often started with *playing* computer games. As some of my interviewees pointed out, such play should not be trivialized. "In the end video games are programmed," explained Célio when talking about playing with his Amiga at age six. Early experiences playing computer games lay a foundation for later computer use. They help children acquire computer skills as well as what Becker (1953) calls "perceptions and judgments of objects and situations" that "make the activity possible and desirable" (235).[2] Becker argues that when a group of people engages in activities in which other people do not engage, this often has to do with the fact that they have learned to see a particular activity and the related objects in a way that makes the activity both attractive and feasible. (Or, to turn it around, those who do not engage in the activity might simply not have had a chance to learn what it is it that makes it attractive or how to go about it in a way that makes it enjoyable or at least bearable.) To be a programmer one must learn to see the computer as an object that can be controlled through an understanding of its inner workings. One must also learn to find satisfaction in the acquisition of this control and in the challenges inherent in it.[3] Computer games provide an early situation in which a child can see a computer this way.

Becker stresses that acquisition of "perceptions and judgments" is a social process: one learns them in the process of engaging in the activity together with others. Being in the right social group is what often makes a difference between acquiring the right "perceptions and judgments" and maintaining a long-term engagement, or trying the activity briefly and giving up. For many of the nerds I interviewed, their earliest interactions with computers were stimulated by interactions with their fathers (or sometimes other male relatives), who either introduced their sons to computers they used themselves or bought computers for their sons seeing it as something that would be worth learning. The long-term and more serious engagement, however, often depended on finding a group of peers and mentors.

Jason found such a group when his father decided to put him in a computer course at age fourteen. This brought Jason in contact with people who would help him develop his interest in computers further:

Jason: Instead of going to a mall we were *hanging around* [says in English] at this computer course. The instructors of the course were experienced people, experienced professionals; they knew a lot, they were good, and so we would be there, picking up tricks and tips from them. People who programmed at a very low level [working directly with the hardware]. One guy knew assembler, another one knew C++, another knew C or I don't know. [. . .] This group of people, we "traded cards" [*trocava figurinhas*], right? We would say: "But how did you manage to do this?" "Ah, I figured out that at such and such interrupt of DOS you can put this thingie and the cursor would then notify you every time that it's . . . you can intercept the pause at the clock and then you can get the key of the thingie and then you can call this program on top of that one . . ." Cool ways to do stuff.

While access to mentors and peers was an important means of learning about computers, we must note that Datacenter also gave Jason access to a milieu in which learning about computers would be understood as *cool*, and where an exchange of findings could be integrated with simply "hanging around" with friends—an alternative to going to the mall, as Jason pointed out. "Trading cards" (roughly equivalent to "comparing notes" in English, but with a more playful connotation) provided Jason and his peers not only with an opportunity to learn from others, but also with a *reason* for learning new tricks.

This social side of Jason's experience should not distract us from the more mundane side of working with software, and the individual effort involved in "pushing horizons" by trial and error:

Jason: Those were difficult times, I remember, because finding information was difficult. To figure out how to do something you had to go by trial and error. [. . .] So, a solution normally was to get programs in whatever way possible, someone who had it would make a copy, and you would go and try checking it out and discovering how it worked. Then you would use its resources, and perhaps find someone who had already done something more advanced with this: "Hey, how did you do that?" Then the guy would explain it to you and you would apply it in your program.

He returned to this topic later in the interview:

Jason: So it took a lot of time to push our horizons. In return, this was very *thorough* [*bem* thorough, *bem minucioso*]. We managed to do things that sometimes surprised the instructors: "Wow, how did you manage *that*?" "Yeah, I had to map all the interrupts there and find out that this one did this, the other one did that. I had to find some way to work around this thing that I couldn't do." That happened . . .

To explain the need for the hard work of understanding the system's low-level behavior by systematically mapping it out ("mapping interrupts"), Jason pointed to the difficulty of obtaining foreign books, which were "crazy expensive" and took months to arrive. This specific problem is rarely mentioned by those who started learning programming later, in the age of Google. One important aspect of software work has remained unchanged, however: now as then, software development requires countless hours of individual work, much of which goes toward understanding why a technical system does what it does and how it could be made to behave differently.

The two sides of software work—the solitary investigation and the social "hanging around"—are inherently linked. Programmers usually understand software work as being, at its best, a process of making discoveries ("cool ways to do stuff") and sharing them. This sharing helps expedite individual discovery work and creates an audience for "war stories" (Orr 1996) about the achieved results. To be able to share, however, one must first discover something. And as many programmers point out, the time that one has to spend alone in front of the computer for this turns away all but those who enjoy this process for its own sake. The effort of "mapping interrupts" requires dedication that is seen as obsessive by outsiders, and often by the programmers themselves, who often say that anyone who is not "obsessed" in this way and does not find joy in this painstaking pursuit of obscure knowledge is likely to find this work too frustrating. "In return, this was very *thorough*," said Jason. Being "thorough" (Jason used an English word here) is its own reward—a *return* for the hours spent with the machine. It is only in the right group of peers, however, that Jason would come to see "intercepting the pause at the clock" as something "cool," a legitimate form of "hanging around," and a reasonable alternative to going to the mall.

"In a Place So Far Away"

While having to order books from abroad highlighted the foreign nature of the practice he was starting to engage with, the power of remote centers over the local practice was not as apparent to Jason in the late 1980s as it is today.

Jason: We wanted to make applications because at that time there were few applications. There were few things. So, since we understood a bit of programming, we thought that we had what we needed to build those applications and become rich and famous. And it was even more exciting to see that we could build things that were good.

Jason then turned to a story about his friend Rogerio, who grew tired of WordStar, a text editor he was using, and decided to write his own.

Jason: So he stated to write a text editor that started to have functionality that was better than WordStar. A 16-year-old kid, stuck [*enfurnado*] in a place so far away! And that was cool, this joy . . .
Yuri: Far away where?
Jason: In Nova Iguaçu, far away from . . . Even far away from the closest metropolitan center, which was Rio de Janeiro, but also far from the place where commercial software was made, which is there in the United States, there in *Silicon Valley* [says in English], et cetera. So, in a peripheral city in a third world country, the guy managed to make a program that in comparison to the commercial software that was available. . . you could say: "This software is *good*!" This potential motivated us to study, to learn things.

As Rogerio was taking on WordStar, Jason himself focused an even more ambitious task: running several programs in different windows.

Jason: I wanted to do something that would allow you to run several programs at the same time in different windows. Now you see: I wanted to do this in graphic form, on the DOS screen, but it was very slow, not very good. I wanted to keep trying better solutions, to put smarter video drivers, to copy the data faster. So I arrived at the conclusion that to do this I would have to use the disk and that it would end up being very slow, so I decided that this would not work and gave up. And went to pursue other things. I was quite annoyed when Windows came out a few years later, using of course the disk—which was the idea that I had and discarded as undoable. I thought: "Damn, if I had pursued this, I would have become rich." [Laughs.] Or not, right? [Long pause.]

Jason followed the story with a long pause, giving both of us a chance to contemplate what would have happened if he were successful in his endeavor.

Jason referred to Nova Iguaçu as "a peripheral city in a third world country"—far even from Rio, not to mention Silicon Valley, the mecca of the software world. He also described the period as a difficult time, as we saw in the previous section. This isolation, however, allowed Jason and his friends to dream big. While Jason remembered Windows coming out a few years after his own attempts to do the same thing, the first version was actually released in 1985, when Jason was ten years old. Jason did not see Windows until 1990. Jason's friend Rogerio similarly focused his efforts on writing an alternative to WordStar—at the time when WordStar was dramatically losing market share in the United States, suffering devastating competition from WordPerfect and Microsoft Word.[4] Technical news took time to reach Nova Iguaçu in the 1980s.

As a teenager, Jason thought he would have become rich had he managed to develop a good way of running programs in multiple windows (something that had made Bill Gates wealthy three years earlier). Now he seems to doubt that this would have helped. While this loss of optimism undoubtedly has much to do with growing up, younger developers rarely express the same sense of excitement as those who entered computing in the 1980s. The Internet has made Brazilian developers simultaneously more and less isolated. While being more connected, Brazilian developers today appear to be more aware of how isolated they are. In 1989, Silicon Valley was a rather vague idea. It was hard to imagine concretely what it would be like to be there. Today, the developers are a lot more exposed to what is happening in the United States. They are thus more aware of being "stuck" in Brazil.

The relative scarcity of foreign applications in the 1980s created opportunities for local developers—at least in the developers' imagination, but also to some extent in practice. This situation changed rapidly in the 1990s, as American companies increasingly began to enter the Brazilian market. Local application developers found themselves being judged by standards that were increasingly difficult to meet. While many opportunities remained in the local IT services market (where local companies get an advantage from their closer relationships with their institutional customers), developing noncustomized software products has been broadly accepted as the domain of the Americans—or, perhaps, the Indians they employ.

People who start product companies are crazy, said Rodrigo Miranda as we sat down to discuss the history of Nas Nuvens, a few weeks prior to my interview with Jason. *João is crazy in this sense,* he continued. *But he founded Nas Nuvens in 1997. Starting a product company now would be even crazier.* In the 1990s the customers knew little of what was happening abroad, explained Rodrigo. Now they compare everything with foreign alternatives. *Then you could say: "This is a search engine." And they would say: "Okay." Now they respond: "This isn't a search engine. Google is. Is this as good as Google? Does it do the same things?"* As we will see in later chapters, successful companies avoid such competition by building custom software for specific clients, where their location becomes a source of strength vis-à-vis foreign competitors.

Rodrigo's memories of the 1980s, however, differed from Jason's. In particular, they did not include a shortage of software applications. Hardware was always hard to get in Brazil, but software was usually easy, Rodrigo told me. By 1990s, there was "the blue box," he explained, a software application that made the computer emit sounds that tricked the phone network into letting you make free phone calls, even international. Some people Rodrigo knew used this application to connect to computers in Sweden for five days at a time to download software. And once one of Rodrigo's friends had the software, they all had it. "We were used to having new software as soon as one week after it was released," Rodrigo added later. He told me of a particular machine in Rio that collected the larger downloads. *Of course its location and its very existence were secret,* he noted. I asked Rodrigo how he knew about this "secret" machine. He smiled. *Obviously all the nerds knew each other,* he explained. Local ties were (and remain) important. Living in the upscale Zona Sul, Rodrigo knew the right people. Fifty kilometers away, in the poorer Nova Iguaçu, Jason was attending the wrong high school.

In either case, however, the newcomers quickly discovered that they were entering a world centered somewhere far away (even if they were not sure how far) and that success in this world would depend crucially on their ability to build links to those foreign centers. In the very least, they had to obtain access to foreign technology—the hardware and the programs. They had to find the books and learn to read them in English. They also had to learn to build local ties, to make construction of global links a collective project.

A Nerd in Transition

After Datacenter, Zé Luís opted for a vocational secondary education, still in Nova Iguaçu, now thinking of pursuing a career in information technology. There he soon found himself ahead of the class. "Secondary school was a *piece of cake*," he said, using English for the last phrase. During his first year in the school, however, Jason started working on his school's database. "Which ended up giving me a lot of access to people, the teachers, the labs," Jason explained. Spending all of his free time in the computer lab Jason could make progress in his learning despite the lack of challenging coursework.

Time spent in the lab also earned him his English nickname "Jason."

Jason: I had access to the labs to do this [work on the database] and this gave me the nickname that I use until today professionally. They called me "Jason," since there was that film "Friday the 13th," about Jason with a mask, etcetera, who would never die. You could shoot him, and then . . . And I was someone really obsessed with programming, so I would go there and program, spending days there.

One day he happened to be free from classes in the morning, having finished an exam early. This allowed him to go to the lab and spend his entire day there.

Jason: I got there and sat there programming, and the groups were coming and going, coming and going, and I stayed there from eight in the morning until eleven at night without getting up from the chair. And since secondary school is a place where rumors spread naturally, the next day the whole school knew of the boy who had stayed at the computer from eight in the morning until ten at night. So during the next week they started calling me all sorts of names: "zombie," "vampire," "the living dead," "without signs of life." What stuck in the end was "Jason." So, everyone would be like: "Ah, Jason who wouldn't die, who is there at the computer, as always."

The story illustrates the tension faced by many young nerds. Jason's obsession with computers was not understood by his peers at the new school (despite the school's technical focus) and marked him as different from fellow students, a theme that comes up in many other biographies. At the same time, however, Jason's computer skills, obtained through hours of "mapping interrupts," began to give him privileged access to important resources. He was also starting to get financial rewards for his knowledge, receiving a stipend in return for his work. A year later Jason got hired as a teaching assistant in his school's computer course and his technical knowledge started bringing him certain rewards in terms of social status as well, putting him in a position of power over fellow students. While intended as a term of ridicule by his classmates, the name "Jason" thus also came to signify his entry into the world of professional software. Today Jason wears this moniker with pride, using it as his "professional name."

Being a nerd is not a career choice, but a way of life often accepted in childhood. It involves, among other things, collective exploration of systems of knowledge that lie outside of the mainstream culture and can range from the imaginary worlds of role-playing games and graphic novels to the interrupts of DOS. Some of those systems of knowledge, however, underlie what Giddens (1991) calls "expert systems"—sociotechnical systems that are essential for the functioning of the modern society yet are opaque to most of its members. People who master "expert systems" (the "experts") face good opportunities for gainful employment. The system of knowledge related to getting computers to perform various tasks supports one of the Giddensian "expert systems" and appears to be particularly appealing to nerds.

As young nerds grow up, they come to realize (often with some nudging by adults) that some systems of knowledge bring more financial opportunities than others and start focusing on them as their future profession, leaving computer games and RPGs as a hobby (and eventually abandoning them altogether). As they move into professional software development, they bring with them some of the skills they acquired in other nerdy pursuits—such as Mauricio's networking skills perfected for the sake of playing *Doom*. More important, they bring with them a set of "perceptions and judgments" that make it possible to see software development as enjoyable because of the opportunity to "push horizons" that it offers.

The transition from software as a childhood passion to software as a profession, however, involves more than a choice of one system of knowledge out of a few attractive alternatives. The future software developer must start to engage in an entirely new way with the social world surrounding computer knowledge. As I argued in chapter 1, a practice such as software development must be understood as simultaneously a culture and a system of economic relationships. A sixteen-year-old nerd who has acquired a good quantity of the culture of software development may nonetheless be quite new to its economic side, and in fact may have to relearn some of the culture. This transformation typically requires a combination of two factors: experience working with people who practice software professionally (typically inside an organization that produces software for commercial use), and the acquisition of a certain theoretical base. Most developers acquire such a theoretical base at least in part in college.

Universities

While the developers are nearly unanimous in stressing their ability to learn by themselves and are frequently critical of their undergraduate experience, most of those who spent enough time in college recognize at least some value in the experience beyond the certification demanded by the employers. For those who attended college full-time, the early years often provide a crucial socialization experience, introducing the future software professionals to a much-expanded circle of like-minded peers (who often form the core of their future professional networks) as well as to people who have engaged with software for much longer. Many also stress the importance of the curriculum and the discipline demanded by some of the university programs. "When you go to the university you have a defined curriculum that you have to go through, whether you like it or not," explained Célio, who attended a public university. He illustrated this point with the story of how he learned Java in 1997, at the time when few people at his university even knew of the existence of this new programming language. Célio had to rely on books and the Internet to learn about Java, but he credits the university for pushing him to learn by requiring him to write a report about a less known programming language. Such learning of course could and does happen in the workplace. Few employers, however, want to train their employees from scratch, and most usually require at least a year of college instruction even for those who have learned programming in high school.

Rio residents usually group the metropolitan area's educational options into three categories. At the apex of the system (at least in computer science) stands the Pontifical Catholic University of Rio de Janeiro, also known as "PUC-Rio" or simply "PUC" within the city. Like other Catholic universities, PUC is private in the sense that it is not run by the government, but the term "private" is rarely used in Brazil when referring to such schools.

The term "Catholic" (*universidades católicas*) is used instead. PUC is highly prestigious. It is also quite expensive. Closely following PUC in prestige— and in some fields surpassing it— are several "public universities" (*universidades públicas*) run by the federal or state government. Such universities do not charge tuition but are very selective—so much so that public secondary education rarely prepares a student for the entrance exam into such schools. (So, perversely, up until recently, access to free public education required prior access to expensive private high schools or private preparation courses.[5]) Additionally, public universities typically have their classes during the day, creating difficulties for students who must work to support themselves, making such universities most attractive to students who can rely on parents to pay for their living expenses. Lower-middle-class students who cannot attend public universities but want higher education are served by "private universities" (*universidades particulares*), which are run independently from the government, usually as for-profit entities though sometimes by nonprofit foundations. Such schools charge tuition, but they accept students with less preparation and offer night classes in a range of convenient locations. Many of them operate on an impressive scale: one runs fifty-seven campuses with a total enrollment of nearly 200,000 students. It is usually understood, however, that quality of education is not one of the strengths of such institutions.

Like many of my other interviewees who attended public high schools, Jason did not see public universities or PUC as an option. Instead, he started a nighttime program at a local private university. He abandoned the program after three months, concluding that he was not going to learn anything useful in it:

Jason: I only spent three months in that university. [. . .] At the time I was a very technical guy, not very . . . I was very much a "bit twiddler" [*escovador de bit*] is what we called it, right. A really low-level guy. [. . .] And so I underestimated such knowledge as high-level analysis. I even considered databases trivial: "Meh, you just take the data, put it there, and later take out, put in again and take it out, put in, take out. Nothing special." So I somewhat underestimated the courses that they had. I kind of evaluated the curriculum from the point of view of the technologies. I thought: "Whatever. They won't teach me any new technology here, so I won't stay here." And so I went home to study other things. And I ended up doing a mix of things: studied this graphics thing, studied programming languages, learned C++ at that time, between '91 and '94, and other things. [. . .] At that time there already were decent books, right. It was already '91, '92. It was possible to go to a bookstore. They opened one, actually here, downtown, it was our *playground* [says in English]. "Livraria Ciência Moderna," here in Edifício Avenida Central . . .

Having studied largely by himself and with his peers up to that point, Jason had appreciation for certain types of knowledge, which did not include the theoretical and "high-level" knowledge that the university program was offering him. (High-level in this case means looking at the overall design of information systems rather than focusing on details.) While he later realized that such topics were important for his work, at the time a small local university lacked the prestige that might have convinced Jason to put aside his own reservations and take the courses more seriously. The seeming ease of learning everything at home, using foreign books that were becoming increasingly available, further contributed to his doubts about the university.

Jason's experience with the university illustrates a more general pattern that we will see many times later: peripheral actors often lack trust in each other, which leads them to focus on direct global links rather than make use of what other local actors can offer. Students assume that local universities (especially the less prestigious ones) cannot teach them anything useful. University instructors often have equally little trust in students' ability to learn.

Jason's quick abandonment of the university program is somewhat atypical, enabled in part by Jason's success in finding a programming job prior to attending university. In contrast, most software developers I interviewed in Rio de Janeiro had completed at least one or two years of a university program in computer science or *informática* (IT) before being able to get any serious software job. The percentage of them who actually complete their degrees, however, appears to be somewhat lower than it would be in the United States.

Those who can get into (and afford) PUC or one of the public schools typically spend their first year or two just studying (and often remember that year fondly as a time of learning), but then look for an "internship" (*estágio*), which often means a relatively permanent job with somewhat reduced hours and additional flexibility, at a reduced salary. They typically finish their program (which usually leads to a raise), but often talk about the last few years of school without enthusiasm. The most important effect of the university program, therefore, is that it helps the future developers obtain an internship.

Those who cannot attend public schools or PUC typically start working in a different (though often related) occupation—for example, offering technical support while simultaneously attending night classes. After a year or two of night classes they can often get an internship, though at a lower salary than that of PUC students. While they often doubt the quality of instruction in the universities they attend, they typically stay enrolled, knowing that such university enrollment will be seen favorably by future employers and that the eventual "piece of paper" will further expand their employment options. Their lack of trust in the program is thus primarily expressed in their lack of attention to the course. After obtaining an internship and eventually more permanent employment, developers who study in private universities at night often slow down their progress substantially, sometimes taking many years to finish the program or giving up altogether.

The Market

Despite his lack of academic credentials, Jason managed to find work, moving through a few small companies and eventually arriving at Petrobras—Brazil's semi-public oil company, widely seen in Brazil as one of the few sites of serious technical innovation. Working for Petrobras had been Jason's "professional dream," but it turned out to be more complicated than he expected, as Jason discovered the limits of the knowledge he had acquired up to that point.

Jason: It was at Petrobras that I kind of started feeling the crisis of my super-technicism. Because I got there and discovered that the world had changed a little, things were easier to do in the world of 1994, 1995, the mid-nineties. The technology had become easier, access to information was easier, and I had a lot of technical background in "bit twiddling" but this wasn't as valued as it was five or six years ago in the late eighties. So I saw a market that needed bank applications, basically information systems, and I didn't have much theoretical background in this, right.

Jason described meeting new kinds of people: "people who came from the information systems from the old days, from COBOL, from databases." Those people were part of a world that existed before, but was hidden from Jason, the world of information systems running on mainframe computers, in many ways also representing a different culture.

This new world gave Jason some credit for his skills as a "bit twiddler," skills acquired from foreign books and long hours spent at the computer—at least enough to hire him. The world of mainframe computing was itself in turmoil, facing the opportunity and the challenge of upgrading to the new microcomputer hardware that was suddenly becoming available—a topic I discuss in chapter 4. It also demanded, however, many skills that Jason did not have. "I was at a disadvantage compared to those guys, because I didn't have theoretical background in data modeling, requirement analysis, things that nobody even talked about at the time." Jason decided to go back to school. His second attempt at a university education, however, ended even quicker than the first. Jason was again disappointed by the quality of instruction and was too busy with work.

Working inside Petrobras also presented other challenges. As is common in Brazil, Jason was employed in Petrobras as a contractor:

Jason: So, it [Petrobras] had this cycle of public competitions for getting new employees, etc., there are people who are employed by Petrobras. But when it needs something done it gets a person from outside, who works on a temporary contract. However, the law doesn't let you do this for long. Labor laws in Brazil are very heavy. So, what does it do? It contracts an external company, a staffing agency, and tells them: "Hire this guy and sell him to me." So the company hires the guy, he does a contract. But then they have to do a tender, right? So they take bids, another company wins, so "Let this guy go. You—hire this guy over there." So the guy is contracted by three different companies, but really he is just working at Petrobras. They want *me* and they invite the companies to bid on contracting me. [Laughs.] I and thousands of people who worked for Petrobras, to make the process faster, to make things work. Because if they were to depend on opening the competition [for employees], that would take time, a year, two, and then there is corruption, what we call "fish soup" [*peixadas*] right—the politicians picking who will or won't get in, they would game the competition, all of this corruption.

A year later, one such reshuffling ("Dismiss him here, contract him over there") resulted in a "bureaucratic accident" leaving Jason without pay for two months. While a few months without a salary would shock few people in Brazil, Jason took it as a sign that Petrobras was not an appropriate work place for a software professional: "As an IT professional I am *naturally* averse to bureaucracy. And when it touches me, I get furious. I got ticked off and decided to leave. So I returned to working by myself." Jason spent the next ten years working for his own company, finding, like many others, that this gave him more opportunity to practice software development in relative isolation from the Brazilian organizational context.

The contractual arrangement that Jason describes appears to be especially common in Petrobras, since its semi-public status means that hiring and firing employees is even more complicated than it is for private companies, but is often used by many other organizations. Many interviewees told me of multiple levels of contracting. While contracting appears to be a lot more prevalent in Rio de Janeiro than it is in Silicon Valley, it is hardly unique to Brazil and cannot be fully explained just by the peculiarities of the Brazilian labor laws (e.g., see Barley and Kunda 2004). Around the same time that Jason endured his arrangement with Petrobras, "permatemp" workers of Microsoft, one of the most "central" sites of software work at the time, were fighting the company over a similar contractual arrangement (*Vizcaino v. Microsoft Corp.*). Yet, the issue is often understood in Brazil as a uniquely Brazilian problem, contrasted with an idealized image of work in the United States. (The US software working environments are often also idealized in other ways. For instance, they are often seen as places where ideas are judged purely on their technical merit, rather than on the personal connections of their originators.) The idealization of the centers illustrates an additional challenge faced by the peripheral participants. When their understanding of how the practice *ought* to work exhibits a clear lack of fit with the local institutional realities, peripheral developers have no easy way of knowing whether the theory carried by the culture of the practice actually fits with the reality elsewhere or simply represents wishful thinking.

Other Paths

Employers' use of passion as a way of identifying "good" software developers suggests that not all developers in Rio de Janeiro start their journey toward this career by falling in love with their dads' computers. And of those who do, few later choose jobs based *only* on whether the work aligns with what they love to do. Finding a job that pays, and pays reliably, is typically a major concern, especially for the older developers who must support their families. Software careers are similar to academic careers in this way. In both cases, the new members are often first drawn to the community of practitioners and its esoteric knowledge. However, those who will continue their engagement with the practice must eventually learn to engage in it in a manner that would allow them to earn a living, freeing them from having to dedicate their time to other kinds of work. To move toward more central (in Lave and Wenger's sense) and more valued forms of participation often similarly requires learning to engage in the practice in very different forms from the ones that may have originally attracted the novice.

The role of passion may also be specific to particular places. My interviewee who stressed his desire to hire developers who "love to program" seemed to have gotten this idea from a programmer essayist working in

Silicon Valley. While this idea resonates in Brazil, it is not universal. In my conversations with software developers in Bangalore, India, including those working for the world's most prestigious IT companies, I typically heard a rather different story of entry into the world of software, one rarely heard in either Brazil or the United States: "passing for computer science." "Passing" means getting a high enough score on the national university entrance exams to get into a computer science program. In India, my interviewees nearly unanimously explained, young people do not *choose* to do software work—they are *chosen* for it. Those who get the highest scores on university entrance exams proceed to study computer science. Those who score less do other things (or perhaps try to get into the IT industry by other means). The outsourcing economy guarantees computer science graduates such high salaries in comparison to everyone else that few would seriously consider not doing computer science when offered a chance. Indian developers often talk about their passion for technology as well, seemingly eager to assuage the American stereotype of them as "mercenaries." Sometimes they stress that they were already science nerds before they encountered computers. However, they have to learn later, in college, what it really means to "love" software.

In this regard, the situation of Brazilian software developers is more similar to that of their American colleagues than their Indian ones. While software development provides good career opportunities, it is one of many upper-middle-class careers in Brazil, and not the best-paying one. For PUC-Rio's Department of Informatics, the most prestigious computer science department in Brazil, attracting good applicants for its day time program in "computational engineering" is a challenge at times, I was told by one of the professors. Those who do well on the entrance exams and can afford an expensive daytime university face many competing options.

PUC's cheaper program in *informática* (information technology), taught at night, is more popular. It appeals to lower-middle-class students who see it as a route to social mobility, though a difficult one. The Brazilian software industry primarily serves domestic clients, who often seek relatively simple systems at low prices. While this may contribute to making the software work less attractive to highly educated Brazilians (who sometimes see themselves as overqualified for the work they get to do), it also creates many opportunities for less sophisticated software work, at lower wages. Many software companies respond to this by hiring developers with incomplete college degrees and occasionally with no college experience at all, then relying on a small number of highly educated individuals to manage and mentor them.

One of my interviewees, "Miguel," started his career at age fourteen as an "office boy" in a software company—an assistant tasked with things like delivering documents to clients, and receiving the minimum salary for the State of Rio de Janeiro, around R\$260 a month.[6] Between the deliveries, Miguel spent time learning to use the computer, and later the basics of web development, relying on conversations with the developers, books, and practice ("reading and testing, reading and testing"). He was eventually allowed to take on simple web development tasks needed by the clients. Two years later Miguel joined his current company as an "intern" working on web development and earning R\$300, while attending high school at night. Another six years later, at the time of our interview in 2007, Miguel was considerably more confident in his skills as a developer, was attending a university at night, and was earning between R\$1,000 and R\$1,500 a month. At age twenty-two he was making substantially more than his father, who had not finished high school. Miguel was looking forward to yet higher earnings in the future. Having grown up in a family that he described as "more towards poor than middle class," Miguel talked about software development in pragmatic terms—a way to make a good living. Some of the other developers I interviewed had moved into software in similar ways.

Miguel's story shows that falling in love with computing in early childhood is not the *only* way to enter the world of software development. The path to software that starts as a childhood hobby is an important one, however. Developers who enter software as Miguel did typically stay at the lower rungs of the software world. This happens for many of reasons. Sometimes they are held back by their lack of theoretical training, sometimes by simple class prejudice on the part of their managers and peers. In many cases they also appear to dedicate less time to software: Miguel had no computer at home. The more ambitious ones also often look for alternative careers, unless they develop a "passion" for software along the way. At the end of my interview with Miguel, I learned of his plans to apply for a government job unrelated to information technology.

* * *

This chapter has explored individual entry into the world of software, looking at the experiences of a small number of people. In many places I have connected such individual experiences to the larger context, touching briefly on topics such as the structure of secondary and higher education in Brazil, the organization of the local software industry, and the history of Brazilian science and technology policy. I have generally avoided treating such topics in depth, however, in most cases limiting my discussion to things known to the participants, who themselves often had, at least at the time, a rather limited understanding of the larger context of their experiences.

In the next chapter, I take a broader and longer look at the world of software, exploring its history and geography, and focusing in particular on how the practice of software development got established in Rio de Janeiro through the combined (though sometimes conflicting) efforts of many different actors. This history will give us a different view of the world that Jason and my other interviewees were entering in the 1980s and 1990s. After that, I return to a more contemporary and local discussion, looking at three specific projects, which represent some of the different ways in which local participants can engage with a global world of practice today.

4 Software Brasileiro

Unlike Mauricio, Jason, Rodrigo, and most of my other interviewees, Ivan da Costa Marques did not grow up playing with computers.[1] The first time he saw a computer was in college, which he entered in 1963. Ivan studied at ITA, an elite technical school located around 300 km away from Rio de Janeiro, which had been established over a decade earlier and was closely modeled on MIT and other US universities. A key center of electronics training and research, ITA was the first Brazilian university to build a computer, and one of the first to receive a computer from abroad. As a Carioca dedicated to spending his summers in Rio de Janeiro, however, Ivan had his first substantial exposure to the world of computing at PUC-Rio's recently established Data Processing Center. Ivan quickly became interested in software and its potential. When Rio's Federal University (UFRJ) established its own Data Processing Center a few years later, Ivan started working there, teaching courses in Fortran and writing software in machine language "just for fun."

A decade later, Ivan came to play an important role in the history of Brazilian computing, becoming a coordinator for the Brazilian government's policy of limiting import of foreign computers in order to create space for local computer makers. The later years of the policy, which became known as "the IT market reserve" (*a reserva de mercado de informática*), cause painful memories to my younger interviewees, who often feel that the policy deprived them of access to proper computing tools in their childhood and youth, requiring them to resort to Brazilian surrogates. My older interviewees provide more nuanced accounts. I will not attempt in this chapter to judge the Brazilian government's policy toward computing technology in the 1970s and 1980s. Instead, I examine the history of the different efforts to establish computing practices in Brazil, placing the market reserve in this larger context.

Looking closely at such efforts will help us understand the constructed nature of the world that Jason, Rodrigo, and their peers entered in the 1980s and 1990s. In particular, it will help us see more clearly the extensive *local work* undertaken to link the world of computing with local and national contexts, and the many choices involved in this process. While my story here focuses on Brazil, I believe a similar tale can be told about many other places. And while the history of the establishment of the software practice in each place has its idiosyncrasies, the result of this process is a remarkable similarity of computing practices around the world. This similarity is, of course, not incidental. After all, the efforts I describe in this chapter did not, for the most part, aim to create "Brazilian" practices of computing and software development. Rather, they aimed to establish global practices *in* Brazil. Even Brazil's closing of its market to companies such as IBM represented, perhaps paradoxically, a globalizing project, as it aimed to bring to Brazil global practices of which, the participants felt, IBM was depriving their country. It is such constant global orientation of nearly all the participating actors, I argue, that ensures that the result of their effort is not a collection of idiosyncratic practices, but rather a set of linkages between the global world of software and specific places—a set of linkages that makes it so easy to think of the world of software as *naturally* placeless.

The establishment of the practice of software development in Brazil and other places cannot be understood in isolation from the larger system of practices related to computing, including the production of hardware and the many uses of computers. It is also important to consider the processes of synchronization that preceded the arrival of the first computers to Brazil, which created in Brazil the context that information technology today takes for granted, from the existence of basic research institutions to the availability of electricity with compatible voltage, frequency, and plugs. Global software has power in Brazil because it is applied in a controlled and *constructed* environment, a "software laboratory," to borrow Latour's metaphor.[2] To describe all the different processes of "enrollment" that went into constructing this laboratory, the story would have to start at least as far back as the beginning of colonization of Brazil in the mid-sixteenth century, if not with the earlier story of the beginning of Portuguese expansion. To keep this chapter to a reasonable length, I focus on the twentieth century, and for the most part events since the 1950s.

This chapter begins with a quick introduction to the history of computing, proceeding from its origins at the two ends of the invisible transatlantic bridge connecting England and the East Coast of the United States and then moving quickly to its modern global spread. I then present several stories that show how the global world of computing established itself in Brazil and in Rio, a particular city at the periphery of that world.

A Global Profession

The ramp-up of World War II in the early 1940s led to rapid innovation in weaponry on both sides of the conflict. This new weaponry required a substantial amount of computation, in particular the production of firing tables for artillery—tables that allowed a gunner to determine the appropriate orientation of a weapon based on its technical characteristics, the estimated location of the target, and the weather conditions. Determining the proper orientation for a particular set of conditions required solving a set of differential equations, and the procedure had to be repeated for hundreds of combinations.[3] In the United States this difficult work was delegated to computers located in Aberdeen, Maryland, halfway between Washington, D.C., and Philadelphia. As the amount of work grew, additional computers were hired at the nearby University of Pennsylvania, with the total number of computers exceeding a hundred. Those "computers" were people—professional mathematicians (usually women) who performed the calculations with the assistance of mechanical calculators. In 1943, two researchers at the University of Pennsylvania, John Mauchly and J. Presper Eckert, proposed automating the calculations by building a machine using vacuum tubes—an electronic "computer" (Polachek 1997). The result of the project, known as the ENIAC, was completed in late 1945, just a few months after the war ended.

The ENIAC's claim to being the first electronic computer is disputed by a number of other systems built around the same time on both sides of the Atlantic, including the German Z3 and the British Colossus. What perhaps makes the ENIAC the most notable of those machines is the ENIAC team's success in commercializing one of their later computers, the UNIVAC. While the ENIAC was created as a singular instance, only a few years later Mauchly and Eckert's operation (which by that point had been bought by Remington Rand) had set up almost two dozen installations of the UNIVAC I. The first of those machines were installed at the headquarters of the United States Census Office and at the Pentagon, both near Washington, D.C. Later ones were installed all around the United States, with the heaviest concentration in New York, where Remington Rand was also based (Ceruzzi 2003).

After a series of mergers, Remington Rand survives today as a part of Unisys Corporation, still a player in the computing business. By the mid-1950s, however, the term *computer* had become firmly associated with another New York-based company: IBM. IBM's success in selling computers was hardly accidental, since the company had come to dominate the business computing market even before the invention of the electronic computer. It did so through its pioneering use of an earlier generation of information technology: a mechanical device that could accept large stacks of cards that encoded data as a sequence of punched holes and would then quickly add up the numbers encoded in a specified field. Originally developed for the needs of the 1890 United States Census, the device soon found wide use in business (see Austrian 1982). Watching the early success of the UNI-VAC, IBM recognized that the electronic computer provided a powerful (if expensive) alternative to its tabulators. By 1951 IBM was selling its own computer. IBM's familiarity with the practices of *use* of business computing had proved to be a bigger advantage than Remington Rand's head start in the new electronic technology (see Campbell-Kelly 2004). By the early 1960s, IBM came to dominate the computing industry to such an extent that its competitors were often jointly referred to as "the seven dwarfs" (Ceruzzi 2003).

One of the things that distinguished electronic computers from the earlier tabulators and nearly all of the earlier electromechanical computers was their universality. Electronic computers were not built for any particular task. Rather, they could perform a wide range of calculations following a set of instructions. They would load such instructions in the same way as they would load the data on which they performed the calculations: either from a stack of punched cards or from a perforated tape. Such stored instructions were called "plans" or "programs," borrowing a term used in a different sense with the ENIAC.[4] The need for such programs created a need for people who would program computers—"programmers." The earliest programmers were women from the ranks of human "computers," which included the six "ENIAC girls" who handled the configuration of the original ENIAC (Fritz 1996; Light 1999). As men were returning from the war, however, women were being increasingly encouraged to return to the home, and programming started to acquire its distinct modern characteristic as a predominantly male profession.[5]

Since the 1950s programming work has undergone substantial changes. As computer makers quickly learned, their customers often spent as much or more money on programming their computers than they did on the purchase of the computer itself. Over time programs came to be seen as a separate and important component of a computing system, leading to the emergence of the new word "software," coined by analogy with "hardware."

The ranks of programmers steadily expanded: from the six women who programmed the ENIAC in 1945 to around a thousand people working in the United States a decade later.[6] While the early programmers were usually hired by organizations that purchased computers or by vendors, the late 1950s saw the emergence of software contractors—companies that would write software for other organizations (Campbell-Kelly 2004). By the mid-1960s some of these companies started selling software as a *product*—offering the same (or essentially the same) software to multiple clients. While the dramatic success of some such companies has attracted much attention, it is important to remember that most of people who write software today—now mostly known as "software developers"—still write software intended for use either by a particular client or by their own organization.

While the number of people who develop software today is hard to count precisely, as a rough approximation it likely approaches about ten million people worldwide, comprising up to 2 percent of the employed population in the most developed countries.[7] Today's software developers are also distributed quite widely around the globe, though their density varies dramatically. Both of those aspects can be illustrated by figures 4.1–4.3, which show a mapping of IP addresses that have downloaded Lua software libraries and modules from LuaForge.org (a web site maintained by Rodrigo Miranda) in 2007–2009.

The maps show a substantial dispersion of software developers working with (or trying out) Lua. With the notable exception of Africa, most populated regions and most of the world's countries are represented, from Nepal and Bangladesh to Paraguay and Nicaragua. This spread is particularly notable, considering that the map does not represent the totality of software developers, but rather just those interested in Lua. Maps drawn for libraries in other programming languages, however, look quite similar.[8] In other words, software developers not only are spread around the globe but also generally tend to use similar technology.

Qualitative data—for example, my own interviews with software developers in Brazil (and also in California and India) and my observation of their work—confirm this impression of substantial homogeneity of the practice. As we saw in the preceding chapters and will see illustrated again in later chapters, software developers in Brazil develop software using essentially the same tools and techniques. They also share jokes, adages, and cultural references. Some of their jokes and cultural references come from IBM of the 1950s and 1960s. Even more often they reference the "hacking" culture of MIT of the 1960s and 1970s. Quite frequently they refer to contemporary software heroes and bloggers. They also share their identity as software

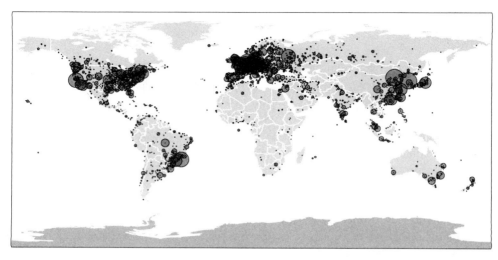

Figure 4.1
LuaForge.org downloads, January 2007–April 2009. *Notes:* The graph was constructed by taking unique IP addresses that have initiated downloads of Lua libraries
from LuaForge.org between January 1, 2007, and April 28, 2009, a total of around
200,000 IP addresses. Those addresses were then mapped to latitude and longitude
using the GeoLite City database. The resulting observed locations were then grouped
with others within 100km of them. The area of each circle is proportional to the
number of IPs at the location, with the smallest circle representing one IP each and
the largest ones representing around two thousand. The GeoLite City database is
provided by MaxMind and is described at http://www.maxmind.com/app/geolitec-
ity. The database maps an IP address to the correct country in 99.5 percent of the
cases and usually places it within 25 km from the actual location (e.g., 79 percent of
the cases for the United States, 54 percent for Brazil). Some addresses, however, may
be mapped to a location that is in the right country but more than 25 km away
from the actual location (18 percent of the cases for the United States, 25 percent
for Brazil) or not be mapped to a specific location beyond the country (3 percent for
the United States, 21 percent for Brazil). Locations mapped at the level of a country
are represented by a circle in the middle of that country.

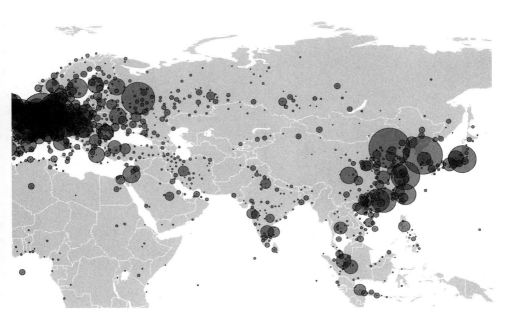

Figure 4.2
LuaForge.org downloads, January 2007–April 2009, Asia. (See notes for figure 4.1.)

developers, nerds, people interested in information technology, often treating this identification with a global community of software developers as a natural part of who they are and as an explanation for their actions, as we saw in the previous chapter.

Whether we think of software in purely technical or in cultural terms, one cannot talk about "Brazilian software" in the same way as "Brazilian music"—that is, as a distinct *kind* of software or a different kind of software practice that could be meaningfully differentiated from "American software" or "Indian software." The title of this chapter is an intentional misnomer. The term "software brasileiro" is rarely used outside presentations by government agencies and industry associations. Most of the Brazilian practitioners I interviewed had little interest in *Brazilian* software development, but were instead keen on expanding and improving in Brazil the practice of software development as understood globally.

While recognizing the global nature of the software practice, however, we must also note the extent to which the practice of software development is established in different places. The map discussed earlier illustrates this variation. Some areas (the two coasts of North America and most of Europe) are covered thoroughly. In other regions, the visitors appear either

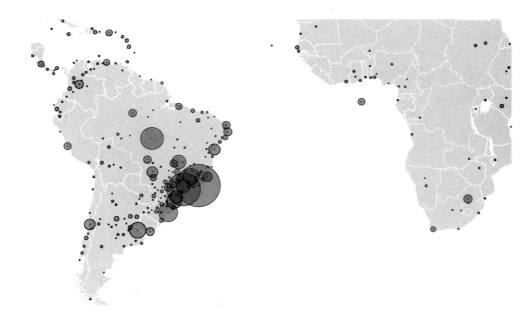

Figure 4.3
LuaForge.org downloads, January 2007–April 2009, South America. (See notes for figure 4.1.)

in smaller clusters or individually. Note that while the map represents traffic to a site hosted in Brazil, the visitors (represented by the points on the map) come predominantly from Europe and North America. Maps of people interested in other software technologies demonstrate similar patterns.

Official statistics confirm the impression that software developers are concentrated in specific places. In the United States, the number of computer professionals can be estimated at around 3 million people (about 1 percent of the population and 2 percent of the employed). Brazil likely has about one-twentieth as much (and one-tenth as much per capita). In Rio de Janeiro computer professionals account for about 20,000 residents of the city, roughly one-tenth of the number of computer professionals working in the San Francisco Bay Area (which has a somewhat smaller total population).[9]

This already substantial difference is amplified tremendously if instead of simply counting people, we look at the nature of their work. There is a general impression among software developers in Brazil that the most important software employers are based in specific places, such as Silicon

Valley. A look at the association between market capitalization and location confirms this in the starkest way. In the spring of 2008 the value of publicly traded "software development" and "computer services" companies headquartered in the San Francisco Bay Area added up to nearly half a trillion dollars, over 37 percent of the total valuation of public companies in those categories traded on the US markets (which includes most of the public non-US companies), corresponding to about $2.2 million per computer professional employed in the area. With another 23 percent of the valuation attributable to the second metropolitan area (Seattle, $4.3 million per computer professional), the software and computer services companies in the rest of the world added up to less than 40 percent of the total. The software industry in a place like Rio de Janeiro is tiny in comparison with the larger centers of the software world. The only publicly traded company based in Rio and engaged in this sector at the time had a market capitalization of about half a billion dollars—about one-thousandth as much as the San Francisco Bay Area's share and roughly the price of a handful of "average" venture-backed companies in the United States.[10]

Perhaps even more important, software platforms used by software developers worldwide also come from a small number of places. Software developers in Rio de Janeiro, for example, work primarily with two operating systems: Microsoft Windows and Linux. While the former is unambiguously associated with a specific place,[11] Linux is often described as a globally distributed project, including, in fact, a number of prominent contributors from Brazil. A mapping of the addresses included with the names of people credited in a 2007 Linux release, shown in figure 4.4, however, again points to a substantial centralization, with the largest dot on the map again appearing on the West Coast of the United States.[12] Other kinds of software platforms—for example, databases and programming languages, whether proprietary or open source—are similarly associated primarily with a small set of places.

To explain this centralization, one could point to the concentration of venture capital and investors' reluctance to put their money in remote companies, especially those located far from the established centers (e.g., Zook 2002; Powell et al. 2002). We could wonder if places like Rio de Janeiro just lack sufficiently smart people, perhaps looking for the flaws in their education systems or at their loss of smart people to "brain drain."[13] We could investigate whether the governments of those places inhibit formation of new ventures through unnecessary regulation.

It would be wrong, however, to stress any one factor as responsible for the concentration of the software practice. Reproduction of practice

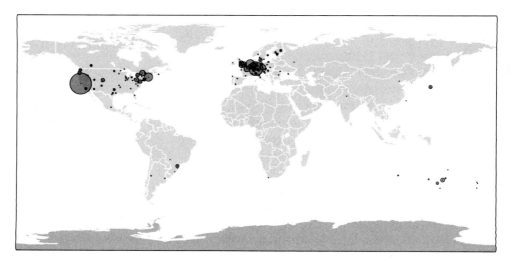

Figure 4.4
The location of the Linux contributors credited in the 2007 release. The map shows the location of addresses included in the credits file for Linux 2.6.22.8. The addresses within 100km are merged and represented by a single larger circle (the area is proportional to the number of people included in the location).

involves re-creation of a system of relationships between many elements of the practice, most of which are mobile only to a limited extent. People who are unable or unwilling to move to the places where the practice is well established must re-create it locally piece by piece, importing some of the elements and seeking local substitutes for the others. The many challenges involved in this process of reassembly are explored throughout the book. The difficulty of re-creating in full a complex practice ensures that new central sites rise to prominence only infrequently. In most cases, it is the decline of formerly central sites and the rise of new ones (for example, the rise of Silicon Valley as the rival of the Boston–Washington corridor) that poses a puzzle rather than their stability.[14]

Even simply establishing a peripheral site typically requires extended work by many actors, who must unite their resources. This process can be complicated by the fact that the local actors must often make a difficult choice between trusting other local practitioners to deliver some of the elements of the practice and attempting to import those elements directly from the centers. The rest of this chapter explores the history of such efforts in Brazil, and more specifically in Rio de Janeiro.[15]

A Brazilian MIT

In 1930 Getúlio Vargas, the governor of Brazil's southern state of Rio Grande do Sul, led a march on Rio de Janeiro, ending Brazil's "Old Republic" and starting a new era in Brazilian history. Vargas brought together a coalition of forces, known as the Liberal Alliance, which included as one of its key constituencies "the Lieutenants," a movement of lower-ranked army officers seeking a wide range of progressive reforms. Among the Lieutenants was Casimiro Montenegro, an army pilot and an aviation enthusiast. Once the revolution was won, Montenegro turned his efforts to creating an airmail network and otherwise popularizing aviation (Morais 2006).

A decade later, as World War II was escalating in Europe, the United States was seeking new allies, including those in Latin America. After some wavering between the Allies and the Axis, Vargas entered a military alliance with the United States, using it to strengthen both Brazil's industrial position and its military power by drawing on American technology. In 1943, Montenegro went to the United States to negotiate a purchase of airplanes for Brazil. While there, he visited Boston and was taken on a tour of MIT by a Brazilian aviator who used to work under his command and who was at that point studying aeronautic engineering at MIT. The visit inspired Montenegro to start planning a "Brazilian MIT"—a higher education institution focused on aeronautics, modeled closely on its American counterpart, and built with the support of MIT (Botelho 1999; Morais 2006). As Montenegro saw it, the future success of aviation in Brazil required not just pilots and airstrips but also local aviation engineers.

Two years later, Richard Smith, a professor of aeronautic engineering at MIT, visited Rio de Janeiro, presenting what became known as the "Smith Plan." The plan involved creation of a large research center, as well as a semi-civilian educational institution associated with it, which became known as the "Technological Institute of Aeronautics" or "ITA." Smith insisted that ITA closely follow the American model, from guarantees of academic freedom to its faculty and students to the use of a *campus* as a model of spatial organization of the institute. Montenegro, who was appointed as the head of ITA, embraced Smith's position enthusiastically (Morais 2006). The first group of ITA students started classes in the early 1950s, first in Rio de Janeiro, but soon moving to the new *campus* in São José dos Campos, a small town located three hundred kilometers away from Rio and hundred kilometers away from São Paulo—a location intended to isolate ITA from Brazilian politics, which Montenegro feared would interfere with the reproduction of the American system of education. Most of the faculty—"the Wallauscheks, the

Theodorensens and the Schrenks" as Morais (2006) calls them—were then contracted and brought to this enclave from MIT.

Around 1960, a group of ITA students, led by one of the imported professors, took a tour of Europe, which included a visit to Bull, the French computer maker. Upon return, the students proceeded to build "Zezinho"—a machine that became known as the first computer ever built in Brazil. While earning a place in the annals of Brazilian history and indicative of ITA's position in Brazil's education system, Zezinho was disassembled soon after being constructed, its parts reused for other electronics projects. The team that created Zezinho was disassembled in much the same way. One of its creators left Brazil the same year for a master's and then a PhD at MIT (Dantas 1988), returning to Brazil only years later. Many of his classmates similarly headed north for graduate studies, finding few opportunities to apply their skills in Brazil.

For Brazilian engineers to be able to practice *making* computers and software in Brazil, another set of practices had to be established first: those involving *using* computers. I describe one of the origins of such practices in the following section.

Governing with an Electronic Brain

Starting in 1920, the Brazilian census began using mechanical tabulating machines supplied by IBM. In preparation for the 1960 census, the American suppliers suggested that IBGE, the organization responsible for the census, use an electronic computer for processing census data—as the US Census Bureau had done since 1951. IBGE originally planned to postpone the transition until the 1970 census, but the election of Juscelino Kubitschek in 1956 changed this plan. Running on a modernizing platform, Kubitschek promised to achieve "fifty years in five" in terms of economic and social development. Performing a computerized census became a matter of national pride (Senra 2007).

The new head of IBGE approved a purchase of "an electronic brain"—a UNIVAC 1105, delivered in the beginning of 1960 by Remington Rand.[16] The electronic census, however, turned into the biggest disaster in the history of Brazilian statistics. The machine suffered from all imaginable problems and in 1964 had to be turned off altogether for several months (Freire 1993, 27). The results of the 1960 census were not fully tabulated until fifteen years later, in 1975. While IBGE has never agreed on the cause of the disaster, Freire (1993) and the people I interviewed who worked with IT at IBGE in the 1960s and 1970s typically point to problems that can be grouped into two classes: those inherent in the UNIVAC itself (described as a "fragile" machine) and the local problems specific to the Brazilian context (the lack of parts, the lack of trained personnel, and various organizational problems).

Unlike the later computers that relied solely on transistors, UNIVAC 1105 belonged to the generation of computers that relied, in part, on thousands of vacuum tubes for data processing. It was a massive machine that required quite a bit of energy and powerful air-conditioning (normally pumped through a raised floor). Installing such computers and getting them to work was quite complicated even in the United States. The UNIVAC had a long way to go in becoming disembedded and mobile.

This "fragile" computer was brought by IBGE to a context that was particularly unfriendly toward it. One of my interviewees, for example, talked about IBGE's unfortunate decision to use punch cards made in Brazil. The low-quality paper used for the punch cards left paper fibers in the punch card reader, which then had to be deactivated and cleaned. Parts ordered from the United States were often slow to arrive (Dantas 1988; Freire 1993).

Even more serious was the problem with staffing. IBGE's computing projects created a need for people who could program and operate computers. To address this, IBGE selected a group of Brazilians and sent them to the United States for training. Unfortunately, subsequent gaps in funding (common in Brazil then as they are today) led many of the trained operators to look for other jobs, leaving those who remained to pick up the pieces. As the government soon recognized, taking new people unfamiliar with computers and training them abroad each time would not work as a long-term solution: IBGE needed a broader local market of people trained to operate and program computers. The solution lay in increased cooperation with Brazilian universities, and in particular with PUC-Rio, located in the same city as IBGE. In 1965, PUC-Rio received another computer, in addition to the one it had been given in 1960.[17]

Finally, UNIVAC did not fit well into the turbulent organizational climate of IBGE at the time. While some of the problems were rectified, what emerged in the long term was a solution that put some distance between the Brazilian government and its computers: unable to replicate the necessary organizational climate internally, the Brazilian government routinely outsources many of its IT needs, to companies that are often a lot more similar to their American counterparts than Brazilian government agencies are to theirs. With those adjustments in place, however, the Brazilian government over time made important steps toward becoming a competent user of information technology (see Evans 1995; Tigre 2003), thus providing an important component for the emerging system of practices.

Informática at PUC-Rio

The same year that UNIVAC 1105 was purchased for IBGE, PUC-Rio received a B205 computer, made by Burroughs. B205 was tiny in comparison with the UNIVAC (weighing just about one ton), cost half as much (around US$1.5 million), consumed half as much energy while running (70 kVa, about the same as one thousand incandescent lamps), and had half the memory (around 16kB, enough to store a few pages of text).[18] The machine was administered by the newly established Data Processing Center (CPD), which was staffed almost exclusively by PUC students (Staa 2003).

Arndt von Staa, now a professor at PUC-Rio, joined PUC in 1961 as an undergraduate in mechanical engineering and soon started working at the CPD. There, in 1963, he met Carlos Lucena, the person most often mentioned by many of my interviewees as the pioneer of Brazilian computer science. Lucena himself had started an undergraduate degree in mathematical economics the year before. Many of the senior faculty members in the PUC Department of Informatics today had started their undergraduate degrees at PUC around the same time in fields such as mathematics, economics, or engineering.[19]

In 1965 PUC received a yet smaller computer—the size of a desk—which made it possible to offer the first computing course, based in the recently created Department of Mathematics (Staa 2003). In 1967 yet another computer was bought and several of the students, including Carlos Lucena, spent three months at the University of Waterloo in Canada, following a visit to South America by the head of Waterloo's computing center. This laid the foundation for a link between the yet-to-be established Department of Informatics and Waterloo's Computer Systems Group, which has lasted to this day.

The same year PUC opened its own master's program, in which many of the classes were taught by the students themselves. Staa (2003) describes the strange "bootstrap" phenomena involved in starting a program without certified personnel:

> The most curious things happened, such as for, example, a student defending his master's thesis having as his advisor a "professor" who had not yet defended his. "Bootstrap" phenomena. Without such phenomena, nothing could have been accomplished. (25; my translation)

Staa uses the term "bootstrap" to describe the establishment of the master's program. While this term is often used colloquially, to refer to achieving something without outside help, Staa invokes the technical sense of this term, which originated in computer science in the 1950s, referring to the different solutions to the "chicken and egg" problems involved in starting (or "booting") a computer.

One of the decisions made at PUC in 1967 was the name of the program, which soon became the name of the field in Portuguese. Staa (2003) describes the decision as follows:

The name of the program came after a long discussion, to decide whether we should brazilianize the term *Computer Science* of the Americans or the word *Informatique* of the French. *Informatique* won, as we considered it a more inclusive term. The first neologism of the field was thus born. It was a master's program in which one gave classes to others, and everyone was trying to learn together everything that was new. (25; my translation)

The resulting term "informática" has since established itself as a normal Portuguese word, extending the language to make it appropriate for the discussion of the new practice. Most of my interviewees today use it as a natural part of their language, applying it also to themselves, as in Jason's description of himself as "an *informática* person" (*uma pessoa da informática*). At the time, however, choosing the term was a decision that had yet to be made, one of the many decisions that would eventually help shape the local context. (Over the following decades the meaning of the Portuguese term "informática" has broadened to approach that of the English term "IT," with the term "ciência da computação"—literally "computing science"—becoming the preferred name for computer science as an academic discipline.)

The term *informática* was soon incorporated into the name of a new department: "Departamento de Informática." Many of the professors employed by the new department, including the head of its postgraduate program, had no doctoral degrees, but the situation was soon remedied after a number of them completed doctoral programs abroad, returning to Brazil in the early 1970s. Several years later, the department opened its own doctoral program, granting its first degree in 1979.

National Informatics Policy

By the 1970s the increasing demand for computers made the Brazilian government worried about the growing cost of imported computers, many of which were underutilized, having been acquired for the status they brought to the agencies (Dantas 1988). A small agency called CAPRE was set up in 1972 to rationalize the purchase of computing equipment to avoid wasting precious foreign currency. CAPRE was staffed by representatives of a group that did not exist until a few years prior, called by some authors the "frustrated nationalist *técnicos*" (Evans 1995) or "anti-dependency guerrillas" (Adler 1986, 1987). Those were Brazilian engineers educated in places like ITA and PUC-Rio, some of whom had received postgraduate degrees abroad. While some of them stayed in academia (as, e.g., did Carlos Lucena and Arndt von Staa), those who looked for jobs outside the universities saw few options that they deemed worthy of their skills. In the intellectual climate strongly influenced by Marxist thought and dependency theory (Frank 1966; Dos Santos 1970), some of them perceived this dearth of interesting technical jobs as indicative of Brazil's broader dependence on the United States and internal social problems (see Evans 1995).

With the establishment of CAPRE, the frustrated engineers realized that the organization's mandate could be used as a tool of industrial policy that would aim to create a local computer industry, by introducing restrictions on computer imports and thus "reserving" some of the Brazilian computer market for the local manufacturers. This policy consequently became known as "the market reserve." A number of successful local research projects suggested that building computers locally should be feasible. Ivan da Costa Marques, a graduate of ITA who had recently returned from doing a PhD in Electrical Engineering at Berkeley and was working at UFRJ (Rio's federal university) promoted the idea of building computers in Brazil by pointing to his own group's success in extending the functionality of an IBM computer. Looking outward to the technological developments abroad, Brazilian engineers also saw other signs that there was a window of opportunity for Brazilian technology. The world of computing appeared to be transitioning from the earlier "mainframe" computers to the smaller and cheaper "minicomputers" based on integrated circuits, which brought a promise of renewed competition in the market that until the end of the 1960s was thoroughly dominated by IBM. Minicomputers were also increasingly assembled from parts supplied by a variety of vendors—parts that Brazilian computer makers could in theory order independently and assemble into their own configurations.

The political climate of the day was also in CAPRE's favor. The soaring oil prices had made the Brazilian government increasingly sensitive to spending what was left of its foreign currency on foreign computers. (At the time Brazil imported most of its oil.) Additionally, CAPRE's proposals resonated with the growing concerns by the Brazilian navy about its increasing reliance on foreign computers in its naval vessels—a fact that did not sit well with Brazil's increasingly independent foreign policy. The navy thus also threw its weight behind CAPRE's project.

It is important to recognize that neither CAPRE's engineers, nor the Brazilian navy, nor the Ministry of Planning (CAPRE's head office) were seeking to isolate Brazil from foreign influences. Rather, each group was looking for a way to participate to the fullest extent possible in global practices that they were engaged in and, more generally, to promote the modernization of the country. As is often the case for peripheral actors, this presented all of them with the choice of whether to focus on building relationships with foreign suppliers of the requisite elements of their respective practice or to build local alliances. In the early 1970s, the conditions seemed right for such a local alliance.

In 1976 Ivan was invited to join CAPRE as a coordinator for computer industry policy. Through his efforts, CAPRE implemented a new policy, according to which foreign companies would only be allowed to produce and sell minicomputers in Brazil if they made generous "technology transfer" agreements with Brazilian partners. The largest companies, such as IBM, chose to withdraw (though still providing mainframes), but some of the smaller international companies accepted the deal as a way to enter what would otherwise be IBM's domain. Several national companies arose in the process, later creating a strong lobby for continuation of the policy (Evans 1995). The existence of such companies made possible (or, perhaps, created a reason for) further computerization projects, strengthening the Brazilian government's position as one of the most competent users of information technology among world governments.

At the end of 1970s, forces close to Brazil's new government of General Figueiredo entered the game, allegedly concerned with security of communications used by the Brazilian foreign service and finding CAPRE's work toward creating a national computer industry too slow (Dantas 1988). An investigation by a military committee concluded that CAPRE's focus on computers overlooked the importance of local production of microchips and software. CAPRE was replaced by a new agency, now run by the military, with a mandate to radicalize the policy to achieve local production of those crucial components—tasks that proved to be impossible due to the tremendous economies of scale and network effects associated with the newer generation of technology. Ivan and some of his colleagues, who were no longer welcome in the government, went to work for the national computer industry that they had helped create. This industry had a number of successes. Some of these companies produced computers under a range of "technology transfer" agreements. Some successfully cloned American computers. Jason's first computer, which he described in chapter 3, was produced by Microdigital Eletrônica, based in São Paulo.

The Liberalization

The new agency's more aggressive policy was expressed in the Informatics Law passed in 1984. However, 1984 was also the year when Brazil started a transition toward democracy. The coalition of forces that had led to the market reserve policy, already damaged by the military takeover (Marques 2000, 2003), started to fall apart. As the industrial policy became accountable to Congress, industries that depended on computers, and whose frustration with the inability to buy cheaper foreign technology had grown, found more opportunities to express their opposition. In 1985 the United States threatened Brazil with trade sanctions, responding to the increasing losses that the restrictions brought to American companies (Luzio 1996). This threat further increased the number of Brazilian industries that stood to lose from the continued policy. As part of its negotiation with the United States, Brazil made a commitment to phase out the market reserve by 1992 (Bastos 1994).

The end of the market reserve is sometimes seen as a tragic collapse of an enlightened national policy under the pressure of neoliberal globalization (e.g., Schoonmaker 2002). It is important to remember, however, that the market reserve was itself an alliance in pursuit of globalization and its end signified, above all, a desire on the part of many members of this alliance to seek globalization by other means. As we saw in the history presented in this chapter, many of the actors that have shaped the policy since the 1940s were to a large extent driven by the same goal: finding a way to engage in Brazil in the global practice of their choosing. Brazilian aviators like Casimiro Montenegro were seeking to establish aviation, but found it hard to acquire airplanes and needed local engineers. Brazilian engineers, created through the efforts of people like Montenegro, were looking for a way to try their hand at the most exciting engineering projects of the twentieth century, such as building computers. The Brazilian government was seeking modern ways of measuring and governing its population, acquiring an interest in using computers and needing programmers to program them. As each group pursued its own globalization project and required elements that had to be provided by members of different worlds of practice, they had to decide when to rely on local practitioners and when to import the original elements of the practice. The alliances between the local practitioners of different trades were thus always marriages of convenience. By 1990, as Brazil was looking for change after two decades of oppressive military rule, many were willing to reconsider their alliances. For many of my interviewees, the end of the market reserve was a moment of awakening that they only wish had come earlier.

The opening of the Brazilian market to foreign computers decimated the Brazilian computer industry, but also led to a dramatic expansion of computer use in Brazil. (The causes of this expansion were many, though, and included, among other things, the end of hyperinflation after the success of the Plano Real in 1994.) The end of the market reserve also left Brazil with a substantial number of people who were trained as electronics engineers but now had few opportunities to work on design of hardware. Many of those engineers found that they could transfer their skills to developing software to run on imported hardware.[20] Additionally, some took refuge in local universities, where they started teaching. One of my interviewees, once an electronics engineer, told me:

Jorge: And we, the electronics engineers, we realized that our space was closing. There was no way for electronics to advance in Brazil. So, there were many centers of microelectronics in Brazil, and now there is only *one*—the only guys who were persistent, they continued. They are a kind of intellectual reserve in this area. [. . .] They are still making chips. They make a Java chip now. [. . .] But we here moved to software.

As Jorge saw it, developing software was an easier task than many of the ones he had faced as an electronics engineer. In a similar way, many former computer companies have transformed themselves into software factories.[21]

Around the same time (1988–1990), as a result of complex negotiations, several Brazilian research centers were allowed to establish digital links with BITNET hosts in the United States, thus becoming BITNET gateways for Brazil (Carvalho 2006). In 1992, Rio and Brazil became connected to the Internet, a new computer network that was rapidly growing in popularity around the world.[22] Access to the Internet enabled real-time access to the World Wide Web, transforming the practice of software development. "Then [in the 1980s] if you knew that the person knew about it, you would spend more time trying to talk to him," explains one of my interviewees contrasting his experience before and after the arrival of the Internet, "It's not necessary anymore. You don't need to, actually . . . And again, this is primarily due to the Internet. You can get any kind of information you want on the Internet."[23] Students studying in Brazilian universities could increasingly complement the knowledge of their professors with direct use of foreign technical documentation.

It is worth repeating—as this fact too often appears to be lost on many of my younger interviewees, who are often quick to make unfavorable comparisons between the limited knowledge of their university professors and the wealth of information accessible through the Internet—that the

Internet did not come to Brazil by itself. So easily taken for granted as the basic infrastructure of the modern software practice, access to the Internet is a complex artifact that required both technical and political negotiations. It became possible in Brazil because of the accumulation of technical expertise in Brazilian universities and the Brazilian government who had over time learned to coordinate their globalization projects.

The late 1990s were a turbulent period for Brazilian informatics, a time of change and much uncertainty about what was possible in the future. Such uncertainty led to both fear and wild dreams. By 2005, when I had started my interviews in Rio de Janeiro, the dust had largely settled and many of my interviewees were ready to share with me what they thought was possible in Brazil and what was not. Access to knowledge was easy—through the Internet. The Internet also served as a great source of free software platforms. There was also no shortage of local customers willing to pay people who could convert knowledge and disembedded code found on the Internet into concrete solutions to their globalization needs. On the other hand, access to capital and foreign markets was hard. The bureaucratic hurdles were there to stay. The most reliable path to success appeared to involve finding local clients, building strong relationships with them, then gradually expanding a service business. The chapters that follow explore this and other strategies for pursuing the practice of software development in Rio de Janeiro.

Free / Open Source Software

Before proceeding, however, I must make a note about another important technological development of the late 1990s and early 2000s: the growing popularity of open source software. As I noted in chapter 0, distribution of software on liberal terms goes back to the earliest days of software development. By the 1970s, however, attempts to secure intellectual property in software were becoming quite common. A new intellectual property regime, which the United States introduced in the early 1980s and then quickly forced on other countries, gave further support to the practice of distributing software under increasingly restrictive licenses.[24] By the second half of the 1990s, however, software distributed under liberal terms was experiencing a resurgence, reaping the fruits of the efforts of many people who had struggled through the 1980s and early 1990s to adapt older practices of software sharing to the new intellectual property regime (see Schwarz and Takhteyev 2010). Such software, rebranded in the late 1990s as "open source," has become especially well represented among software platforms, that is, software upon which other software is built. While Windows has remained by far the most popular operating system for casual users of software, a substantial part (and by some counts most) of the deeper layers of the world's IT infrastructure today run on Linux, an open source operating system.[25]

The rise of free software has undoubtedly been a boon to peripheral programmers. In the 1970s, CAPRE's engineers had to fight to get foreign companies to license their technology to Brazilian manufacturers. Today Brazilian programmers, on the other hand, are granted the right to inspect, modify, and redistribute some of the world's most advanced software technology without even having to ask for it. My interviewees take note of this. Some of them express great enthusiasm about it. Others take it as a matter of fact—this just happens to be the way the world of software works today. Some of them use open source software today because they subscribe to its vision, often recognizing such vision as a key element of today's software culture. Others use it because it works well and does not cost any money.

The benefits of free software have also been recognized by actors within Brazil's government, which has pursued, since 2003, a policy of promoting such software for the government's own computing needs (Schoonmaker 2009; Shaw 2011). Software developers I have talked to since 2005 seem to take a somewhat ambivalent attitude toward this policy. Some welcome it in principle, but are doubtful that it would have much effect in the hands of the government bureaucracy. Others are wary of the government having big ideas, regardless of what those ideas might be. (This in many ways reflects the general attitude that Brazil's middle class often seems to take toward government programs.)

Using open source software, however, is not the same as developing it. While some of my interviewees have worked on hobby projects that they released under free software licenses (or, more often, *plan* to release one day, when they have time), most of them spend the majority of their programming time working on proprietary software for money. Some express no discomfort with this fact. Other say that spending time working on open source software would of course be great, but they cannot afford to work for free. Jobs that pay developers to work on interesting open source projects exist in theory, but are hard to find in practice, especially in Rio. Some consider setting up a business around an open source project, but again find this difficult in practice and move on to other things. Alta, the company whose story I tell in the next chapter, provides an example of this. The projects that I explore in chapters 6–8, however, are open source projects, which aim to not only release software under free licenses, but also to engage (in different ways) with remote communities of users and collaborators.

5 Downtown Professionals

It was late March 2007 and I was in a *kombi*, speeding in the direction of "Centro," Rio's commercial district. Taking elevated highways from the campus of Rio's Federal University on Ilha do Fundão, the minivan flew over many of Rio's favelas, finally landing on Avenida Getúlio Vargas, a block-wide avenue, cleared in the mid-twentieth century to modernize the city. I got off at Rua Uruguaiana, a pedestrian street that serves as an entry point to a few remaining blocks of old windy streets, lined with lunch restaurants and office fashion stores, and filled with vendors selling pirated films and counterfeit watches. After two blocks, I arrived at Largo da Carioca, a wide square at the heart of Rio's business district. There I waited for Rodrigo Miranda who was going to take me to "Alta," a successful Java company that, I was hoping, could become one of the sites of my ethnography. Each minute of waiting felt like an hour in Rio's heat, but I knew it was worth the wait. I had spent the previous few weeks trying unsuccessfully to get myself allowed to come and spend a month inside a Java company. Rodrigo's introduction could make all the difference.

When Rodrigo arrived a few minutes later, we headed south, crossing Avenida Rio Branco, and entering a tall building that was all too familiar to me—I had by that point interviewed people from no fewer than three companies in that building. As is typical in such office buildings in Rio, the lobby had a system of "optimized" elevators, each going only to a range of ten floors, some of them with long lines. Joining the longest line we ran into several guys Rodrigo knew; it turned out all of them worked for Alta. We followed them to the office, which Rodrigo entered without introducing himself at the door, as if his presence there was perfectly natural. We paused only briefly in the lobby to appreciate the fancy engraved logo on the glass panel that separated the lobby from a large room.

As we entered the large room, I saw three dozen tables, organized into bays and separated by short dividers. Everyone had exactly the same table—including the owners of the company. All but one person looked under thirty. It seemed like a by-the-book implementation of a Silicon Valley startup from the late 1990s, complete with a beanbag. I followed Rodrigo as he shook hands with people, nodding, making our way toward "Felipe" and "Luís," seated at a desk in the corner. It turned out that they were two of the three cofounders of Alta.

The four of us went to a conference room where Rodrigo introduced me as a Berkeley doctoral student and a *Herculoid.* "Herculoids" was the name of Rodrigo's private mailing list, which he used mostly to forward technology news. When introducing two subscribers who had not met before, Rodrigo almost always introduced them to each other as "Herculoids." ("It's like saying that you are a friend," he explained to me after we left Alta.)

Rodrigo summarized my research project and talked about what an "opportunity" it had been to be interviewed by me and how he strongly recommended that Felipe and Luís agree to be interviewed as well. The two seemed unsure of what to make of me but started talking about their company. They were two of the three cofounders, all recent graduates of PUC-Rio's Computational Engineering program. Felipe was now increasingly doing "the commercial part." Luís was working on a new company inside Alta. The third co-founder, "Eduardo," was Alta's main "technical guy." Earlier they all used to work with any programming language a client asked for, but now they were trying to focus on Java. Ninety-nine percent of their work was now in Java, said Felipe. "99.9 percent," Luís corrected him.

This chapter looks at Alta as a relatively typical context of software work in Rio de Janeiro, providing a background for the later discussion of Lua and Kepler. I use the word "typical" with caution, however. Already in 2007, Alta was a highly successful company that in many ways seemed to have played all of its cards right. When I returned to Alta a year and a half later, I discovered that it had grown substantially, employing close to a hundred people and occupying three floors of a downtown building. It was now responsible for the main user-facing web site of a major Brazilian brand. While Alta was doing what most other Rio software companies were trying to do, it was clearly pursuing this strategy more successfully than many of my 2005 interviewees. I will try in this chapter to point out some of the reasons for Alta's success, but this will not be my focus. Rather, I look at Alta as a successful implementation of a particular approach to peripheral engagement in a global world of practice: bringing global technology to local clients and acting as local representatives of the global world of practice. While operating primarily with "standard" foreign technology, this approach involves an arms-length relationship with foreign centers of the software world, combined with intense engagement with local organizations. Later chapters introduce two other configurations. One of those involves production of technology with global significance locally, but in relative isolation from local needs. Another, a more complex case, involves an attempt to bring into alignment a wide range of resources, both local and global, in production of a global project aligned with local needs.

The Birth of a Software Firm

As I later learned from my interviews with Felipe, Luís, and others, the company started in 2003, four years before my arrival there, when the three cofounders left "Kubix," another company where they had all worked during most of their university years. Unsatisfied with the limited growth opportunities offered by Kubix, the three decided to start their own company, taking advantage of PUC's startup incubator, which provided cheap office space on the PUC campus, free phone and Internet, and help with tasks like marketing.[1] While most software companies in Rio work in software consulting, building custom software for specific clients, entry into the incubator required a business plan that would involve marketing a software *product*. Felipe, Luís, and Eduardo wrote a business plan around commercializing Eduardo's recently completed master's thesis, which proposed a method for integrating enterprise information systems. Two months later Alta opened its doors in the incubator with an initial investment of R$18,000, to which each of the cofounders contributed equally from their Kubix savings.

The product envisioned by Alta's business plan—"InterJ"—never fully materialized, and from my conversations with Alta's founders it is hard to tell exactly how seriously they had taken it. The conventional wisdom in Rio de Janeiro is that a software *product* company cannot survive in the city. ("It's not California," many developers explain.) The incubator, however, wanted to see a product plan and Eduardo's thesis made it possible to tell a believable story. The founders themselves seemed to half-believe the plan, putting an intern to work on polishing Eduardo's code, while debating among themselves whether to sell licenses for the product or to release it as open source and hope to make some money on related services. They decided to go with the open source option, but were not ready to bet on the product's success, if only for the lack of anything to bet: the starting capital of R$18,000 hardly provided a financial foundation for a product launch. Launching a software product requires a substantial upfront investment, regardless of the license under which the product is distributed. While successful open source projects do often end up attracting contributions from a community that forms around them, building such a community is no easy feat and itself requires a substantial investment of effort (as we will see later in chapter 8). Despite the founders' ability to imagine a grand open source future for InterJ, their immediate concerns were with the short-term survival of Alta.

When an opportunity came to do some work for Petrobras, the fledgling entrepreneurs decided to take it.

Felipe: This started right in the beginning, because it was one of Eduardo's contacts. A friend of his who worked at Petrobras needed a company that could do maintenance on this contract, with intranet in this case, and so he asked if we were available. And we were: "Well, of course, let's start earning money and get into Petrobras, which is a large company . . ." And so we started.

The company proceeded to accept a range of service contracts, which involved software development in different languages, training courses, and other tasks.

One of their earliest projects involved Lua. "Fernando," a PUC student who had worked with Felipe and Luís at Kubix, wrote a library (a set of software modules) for linking Lua with Java—as an assignment for a class—and made a presentation about it at PUC. Rodrigo Miranda attended the presentation and offered to pay for some additional features from Kepler's grant. Because Rodrigo's grant did not allow him to pay Fernando directly, Rodrigo made a contract with Alta, which then hired Fernando to work on the desired features. Luís and Eduardo soon joined too, contributing to other Kepler modules as well. Since the company was just starting off, any income was welcome. But it was not so much about the money, stressed the founders. Above all, they wanted to have a good relationship with Rodrigo, who was ten years their senior and knew a lot of people. The investment in personal relations paid off: Rodrigo later matched Alta with its first large project, which the company was later able to use as a success story when making proposals to other clients. The founders remembered the favor. "I think that if we hadn't gotten this project Alta wouldn't even exist today," said Felipe. (The fact that I was allowed to spend time at Alta and write about it perhaps had a lot to do with this as well.)

As the consulting business grew, InterJ was increasingly put aside.

Felipe: So, from the start we moved our focus a little away from investing in InterJ, investing in integration, and from there it just grew. And as time went by, we were working more and more with development of web applications, development of solutions on demand for clients, you know. And InterJ was something we were putting more and more to the side.

When some time later Alta became an official "partner" of "EIT," a large software company based in California, the fate of InterJ was sealed:

Felipe: And a moment came also when we managed to get a partnership with a large, multinational company, which is EIT. We got a partnership with them, and they are one of the main companies that sell integration software. So, we decided to give priority to learning their platform, and to try to offer services on top of it. And this closed the coffin on InterJ. There was a moment when we decided to give priority to working with the technology that is the market leader. Which was EIT. [. . .] Over there, they have developers, in the United States, Indians, all working to improve this system. And we here have nobody. [EIT] was light years ahead.

Alta's founders' stories about InterJ have a touch of nostalgia, though seemingly not so much the painful nostalgia of squashed dreams as the sentimental memories of lost naiveté. The decision paid off, however. Early consulting revenues allowed Alta's founders to hire their most talented friends. The partnership with EIT, a household name in Java circles, became a source of larger and larger contracts. Alta was invariably considered a success by those who knew of it. When I started frequenting Alta's office in June 2007, the company's only problem seemed to be finding place for all the new developers it was hiring. This problem was resolved two months later when Alta got the chance to rent an additional, even shinier, office in the same building.

The Cutting Edge

While more successful than some of its competitors, Alta was hardly original in its strategy, using global technology to solve local problems, providing customized IT solutions based on Java web technology to local clients. (Or, as some of the examples in this chapter suggest, adapting local problems to fit the available global technology.) According to the PowerPoint presentation given to new employees, the company's focus was on "consulting, integration and development of applications based on **new technologies**," with the last words shown in big, bold letters. The founders and employees I talked to at Alta were equally excited about the company's commitment to using the most advanced technology, "technology at the tip" (*tecnologia de ponta*), a term that could be likened to the English "cutting-edge." Another term I heard frequently was "*tecnologia padrão*." While *padrão* literally means "standard," it could perhaps be better translated as "world standard," since it typically connotes the quality rarely found in Brazil, rather than the mediocrity that is sometimes suggested by the word "standard" when used in the United States.

As later slides explained, in 2007 using new technology meant building web applications in Java. Alta was "a 100% Java company," declared yet another slide. In reality, of course, conservative clients and the need to maintain legacy systems often required Alta's engineers to work on somewhat outdated technology. What the company advertised, however, was its *ability* and *willingness* to use the latest solutions. Alta's developers, or, rather "technology professionals," as they were called in the same presentation, followed the latest technology news, often in English, and peppered their speech with portuguesified English phrases, from technical terms and business terms ("essa *delivery*," "o *deploy*") to general phrases ("é *feeling* mesmo").

The company's mission statement, stated in the same PowerPoint presentation, however, presented Alta's other, *local* side: "to transform your desires into reality through IT services tailored to your needs." Rather than selling a uniform product, Alta cultivated durable relationships with local clients, most of whom were quite literally within walking distance from Alta's office. Understanding the clients and their needs was a skill in which Alta's more senior personnel took pride. Alta's relationship with some of its clients was often so strong and durable that the company almost acted the part of an IT department for some of them. "Intermercado," a large Brazilian retail conglomerate and Alta's largest client, retained nearly half of Alta's employees on a *per month* basis, paying a monthly rate regardless of whether there was any work.

In a way, Alta was thus positioned at "the cutting edge" in the sense not only of using the latest global technology, but also of being at the frontier of the global world of software, a point where its cutting edge met the thicket of Brazilian organizational reality. Alta's position often seemed to put it in an ambivalent relationship with its local context. Alta had to convince clients that it was sufficiently local to understand their needs in a way that true outsiders would not. Alta's engineers, however, also needed to maintain their image as outsiders, the representatives of the global world of technology, to help clients believe that their own global dreams would come true with Alta's help. As Alta's engineers talked to me, an insider coming from the heart of Silicon Valley, their position on the frontier of the software world sometimes came across as a recognition of a handicap that they urged me to consider. At times, however, I could sense a feeling of pride in their war stories: pride of surviving and thriving in this hostile climate. I illustrate some of those points in the next section.

Skol with Fabio

A few weeks after my first visit to Alta, I went back to the office to meet "Fabio," one of Alta's young managers whom I had encountered briefly on my first visit. Fabio came downstairs and met me outside as soon as I called him from the lobby, apologizing for being a minute late. He was dressed informally, much like a typical engineer, though with more attention to style. I noticed this mostly because of the stark contrast with the stylish dress shirt that I had seen on him the previous time. As I later learned, wearing clothes that strike a careful balance between dressy and hip was the rule for Fabio, and today's t-shirt was an exception. Perhaps noticing me looking at his t-shirt, Fabio explained that he had a busy day, because they had to deliver a project to a client. He would in fact have to go back later to finish some work, he added.

The space in front of the building was covered with yellow plastic chairs, all decorated with the logo of Skol, a popular beer brand. The number of those chairs, almost all occupied, made me wonder if all of downtown's employees had decided to come out for a drink at the same time. I soon realized that this was precisely the case: it was a Thursday, but Friday was a holiday. "Happy hour," Fabio said in English. We sat down at one of the few free tables; Fabio leaned back in the plastic chair, lit a cigarette, and asked a waiter for a large bottle of Skol, which soon arrived with two plastic cups.

The conversation returned to Fabio's attire. He normally dressed up, he explained, because he was now a minority partner in the firm and was interacting with clients a bit. They needed to make a good impression. Also, he was only in his mid-twenties, and clients tended to dismiss younger people, he explained. So, he had to compensate for that by looking more serious. "More like an adult," he said. Otherwise, he explained, they would think: "What does he know?" Also, he sat in the management section of the room, Fabio continued. They were all together in one room, but there was a management corner. They talked to clients in the conference room, but the clients often wanted to see the office. *Now, imagine*, Fabio said, *that the client walks into the office and sees a guy in torn up clothes, with huge hair, a giant beard. What does the client think? He thinks, "Awesome, this guy is a nerd, he probably knows how to program!"* But when they look at the management corner, Fabio explained, they want to see people who are well dressed, people who look like they would understand their business.

We then talked about how Fabio had ended up at Alta. As was common in my interviews, the story featured a simultaneous immersion in the global culture of software and in local social networks. It also showcased the tensions between the notion of software development as a fun activity and Fabio's emerging understanding of himself as a "professional," who was not only participating in the labor market but also increasingly involved in managing the labor of others. Fabio came to Alta in 2003, from Kubix, where he had worked together with Alta's founders who brought him along to their new venture. Starting as an intern, Fabio earned a little under R$400 a month. Of course, Fabio explained, he was then a *newbie*, a *beginner* (he used two English words to describe his situation at the time.) As he learned more, his salary rose dramatically. A good programmer would not agree to work for less than R$3,000 and could ask for a lot more, Fabio told me, without getting into the details of his own compensation. There are limits on engineers' salaries, however, say most developers, and to increase their income further the most talented engineers typically move to management. Fabio was starting this transition this year. About a year and a half ago, just before he graduated from the university, Fabio had become a minority partner at Alta.

The talk of money seemed to bore Fabio, however. *Let's get back to programming*, he said eventually. I suggested that we talk about Fabio's experience with Lua. Like his friend Fernando and Alta's co-founders Luís and Eduardo, Fabio was a contributor to one of Kepler's projects. He had started working with Lua back at PUC, Fabio explained. He took a class from Roberto Ierusalimschy, and another one from Roberto's spouse. Both had programming assignments in Lua, and in the case of Roberto's class the task was to contribute something of value to the Lua community. While some other PUC graduates I interviewed resented being required to learn a "homegrown" programming language, Fabio took the assignments more positively, since some of his friends at Kubix, including Luís, were fans of Lua. (Luís in turn attributed his interest in using Lua to seeing it used extensively in Tecgraf, a research lab at PUC.) When Fabio later needed a final project for his engineering degree, Rodrigo, whom Fabio had met through Luís, suggested a new version of a Lua programming editor, originally written by Luís and based on Eclipse, a popular software development tool.

After finishing college Fabio continued to work on the tool together with Fernando, getting paid a small amount from one of Rodrigo's grants. (As Fernando told me in a later interview, working on the project was one of the many things he and Fabio did together on the weekends.) It was hard to find time for it, Fabio said, but he wanted to continue participating. He gave me two reasons. When you work with something all the time you get tired of it, he said. He was working with EIT's "eWeb" platform all day, which he also found to be "very commercial." Working on Lua was a diversion. It was not serious commercial technology, he explained, not something he would suggest to a client, but it was *fun*.

It was also about being a part of the Lua community, he added. When he first started using Lua, Fabio explained, he began reading the Lua list and seeing the announcements of people releasing their code. This helped him realize the meaning of "open source." "It's an exchange community," he said, hence the need to contribute. But it was also a matter of personal satisfaction. In any community, said Fabio, be it your samba group or your church, you want to be known, be a member, be someone who has done something for the community. From this, you get satisfaction. When he released his editor, people replied: someone said "Great!" and another guy wrote back and helped correct a mistake in the English used in Fabio's code. So there was feedback, and people were contributing back.

"Okay," I said, "I understand the attractiveness of an open source community, but why Lua?" Fabio's answer surprised me. The Lua community offered him a space to participate, he explained. Not so with Java. Consider Java, he continued. There are Java User Groups, like RioJUG. RioJUG had about a thousand people on the list, but the level of discussion was very low. There were probably three or four people who actually understand the technology and who had good questions. The problem with Rio's Java list, continued Fabio, is that the number of people who understand Java is low, but the number of people who *think* they understand it is high. Why? Because Java is fashionable. The market seeks people who know Java. So the guy gets a book, copies an example, runs it—now he "knows Java." Statistically, continued Fabio, there might be more "Java programmers" in Brazil than in the United States. But the difference is in quality. "Those people here, they don't even know how to use Google!" he exclaimed. "You have to use Google. Google is everyone's daddy. Before asking a question you have to think: someone probably had this problem before, so go look in Google!" That's something people on the Lua list know how to do. And yes, of course you need to know English to make use of Google, otherwise those thousands of results would be of little use. And half of RioJUG people do not even know English! Fabio continued with increasing passion, eventually moving on to Alta's difficulties in finding good Java programmers and the company's need to hire "raw guys" and teach them from scratch.

Neither Fabio's frustration with the local Java community nor his fascination with the Lua community were unique to him. The complaint that many Brazilian "programmers" do not know what they are doing is something I heard quite frequently, not only from the graduates of elite programs like PUC's Department of Informatics but also from those who themselves might be the source of frustration for PUC graduates. (Or, alternatively, they often talked about the incompetence of Brazilian managers or politicians. In either case the complains were often counterbalanced with stories of Brazilians' ingenuity in dealing with problems created by their less capable countrymen.[2])

Fabio's praise for the Lua community is also repeated by everyone familiar with the group. On the Brazilian side, most list members are former graduates of the PUC-Rio Department of Informatics—Brazil's best computer science department. Even they, however, are eclipsed by the list's foreign members who can quickly answer the most complicated software questions. The caliber of the foreign subscribers can probably best be explained through self-selection: in order to discover Lua, still a somewhat obscure language today and virtually unknown until a few years ago, most of them went through dozens of programming languages, driven by curiosity and a search for perfection.

For Brazilian engineers like Fabio, Lua could thus be a ticket into a highly exclusive international (or, one could say, *foreign*) community of developers, and an escape from the mediocrity of local groups like RioJUG. Links to PUC and an early start on Lua give them relatively easy entry into this community. Gaining comparable standing in a different group, e.g., the Linux kernel developers, would be a lot harder. Eric Raymond (1999) talks about the "ergosphere"—the space of work—as one of the key resources in the open source community. According to Raymond, open source communities are gift cultures, where one gains status by offering gifts to the rest of the community. Open source gifts are solutions to technical problems. Much like in academic research, good problems—the ones that would result in valued gifts that are not too costly to produce—are scarce. Ties to Lua's authors and Rodrigo gave PUC graduates like Fabio an opportunity to identify good problems within a small but growing community, or jump into an existing project. The results of their work thus enjoyed ample downloads (one of the key measures of success in open source) and recognition within the community.

This success in the open source community, however, was hard to translate into income in the Brazilian market. If Kepler developed more, it could potentially allow them to make a realistic proposal to their clients, said

Fabio. He was confident that this would happen eventually. He talked about Rodrigo's recent idea to "open" Kepler and invite more participation from people outside (see chapter 8), saying that this would give people more confidence in Kepler. (It would no longer be the "Kepler team" anymore, but rather the "Kepler community" behind the code, explained Fabio, saying both phrases in English.) But for now, he pointed out, there was no "case" of Kepler, no large company using it. And Brazilian clients were not into experimenting. When a client asks "Why use Java?" the answer is "Because giant companies use it." (Fabio rattled off names of a few large US companies.) Compare this with Kepler, said Fabio. "Why use Kepler? Because it's good." This just does not mean much to the clients. They would ask: "Who is Rodrigo Miranda?" I know him, says Fabio, you know him, but they don't. "Who is Márcio? Who is Tiago?" Alta's clients wouldn't know. They knew IBM and SAP.

The distinction between using "good" technology and the technology clients wanted was not limited to Lua. During my time at Alta in 2007, most of the company's work was based on EIT's eWeb. There was a strong sense among Alta engineers that eWeb was no longer the best option, or at least not in all cases. The developers' attention turned increasingly to the many open source alternatives, which offered a number of advantages. First, open source was *cool*. Second, open source solutions were free, leaving the customer to pay a larger portion of its budget to Alta for customization. Finally, such solutions made it easier for the developers to fix the bugs, since the code was open and there was more free documentation online. For those reasons, Alta tried to move its clients to open source solutions whenever possible. With the largest clients, however, Alta did not have this luxury. Such clients typically had a prior relationship with EIT, which sold them eWeb licenses and then offered to recommend a "solution provider"—a local software company that would customize eWeb for the customer's needs. Alta was one such provider, but not the only one. When a client such as Intermercado called Alta saying that EIT had suggested Alta as a company that could implement an eWeb-based system, Alta's managers had to keep to themselves their opinions about advantages and disadvantages of eWeb. Additionally, EIT typically assigned their own software architects to supervise Alta's work and to make sure that Alta was not introducing any open source solutions that would serve as alternatives to upgrading eWeb to a newer version, at a charge. As the client contemplated whether the expensive upgrade was worth it, Alta's engineers worked with outdated technology, something the company would rather not talk about in its presentation for the new employees.

Building an Online Store

A few weeks later I was heading to Alta's kitchenette to get some coffee when I ran into Mauricio, one of the developers working under Fabio's supervision. "Follow me," he said, and led me to the corridor, where Fabio was smoking with a cup of coffee. We exchanged some small talk. "Now let's talk seriously," Fabio then said suddenly. Intermercado people had gotten back to him about the scheduling program that Alta had just delivered to them. They said there had to be a way to set events that would start on one day and continue to the next. He already thought about it and it wasn't so bad. It would require some changes to the database, some to the interface. He started explaining a solution. Mauricio interrupted him: "Why not just have the user enter the beginning and the duration of the event?" "Ah, good idea," Fabio replied. He started thinking out loud, following the suggestion made by Mauricio. This will make it much easier. So, we just let them do this, and then have JavaScript show the end time. Again, the database would need to be changed, as well as some of the user interface. Fabio listed the specific things that would need to be done, walking through the steps that the user would have to go through. He then turned to how long the change would take. "Three days?" he asked Mauricio. "A couple of days to write, a few days to test," Mauricio responded. "Let's ask for five," Fabio summed it up, "Three to write, two to test." Let's ask for five, he explained, so that we can then agree on four, if they insist. Finishing early is okay. Better than asking for four days and then having to accept three. He asked if Mauricio could start right away. Mauricio explained that he had an exam. (It was about 5:30 p.m., and Mauricio was doing a college program at night.) Okay, said Fabio, let's start tomorrow.

As we walked back to the kitchen, I thought about asking Fabio if this was something I could help with. I had been at Alta for a week at this point and was finding it hard to understand the details of what people do without being involved in a project. Fabio anticipated my question. "How much memory does your laptop have?" he asked. "One gigabyte," I said. "That's not enough to run eWeb," he sighed. We agreed that I would help with the user interface, since this could be done without running eWeb. We went to Fabio's desk and he showed me the application they were building, then emailed me the URL and the password. I spent a few hours playing with JavaScript for the new form. The next day, however, Fabio informed me that the issue had turned out to be much more complicated, and in fact the requirements for the project were being reconsidered altogether. Instead, he said I could join him on another project that they were just starting: an e-commerce web site for a different client.

Early next week I was sitting next to Fabio at his desk, watching him draw a diagram that represented the relationships among the "business objects" of a client's online store and was to serve as a blueprint for a database. (At this point we had agreed that I would later help with the database design.) As Fabio explained, they had done design "by hand" in the past, but he now wanted to try doing it with a proper tool. I soon realized Fabio was using not just a diagramming tool but a UML editor—a tool specifically designed for expressing relationships between software objects, which could later be used to generate code and database design.

Fabio was working without too much haste, explaining to me along the way what he was doing. He seemed to be coming up with the various properties of the objects without having to look them up anywhere or even pausing to think. I watched as he created a box for "Product" and then typed in the attributes that a "Product" was supposed to have. He then created another box, labeling it "SKU"—the English abbreviation for "stock keeping unit," he explained. I thought about the "requirements document" that Fabio had sent me the day before, realizing that it did not have nearly enough detail about what the customer wanted. I asked Fabio how he knew what "business objects" the store needed? Fabio laughed. *Those are the same for all stores!* They had been working with online stores for quite some time, he explained. This particular client did not even have anything specific in mind. It had only a vague idea of what a web store would be like. All the client had asked for, explained Fabio, was for Alta to build "an e-commerce web site," and that this web site would be no worse than its competitor's, plus a few extra things.

As I realized later, the request for "a few extra things" did not refer to anything specific either—rather, the client just wanted its web site to have *something* that its competitor's site did not have, leaving it to Fabio to figure out what that something could be. Fabio approached the task by looking at the competitor's site to see what it did and thinking what Alta would need to do to allow for the same features and a little more. He then had a long meeting with the client, explaining what his team could and could not offer. For instance, they agreed that all products would be sold in predetermined quantities: a customer who wanted to buy some *queijo minas* would be able to choose between a 200g or a 300g package, but not anything in between. In this way, Alta seemed not only to be providing global solutions for the clients' problems, but also helping clients adapt their practices to the possibilities of global technology. Fabio continued adding boxes and attributes as he spoke—the task seemed to consume little attention. At one point he paused, to think about what to do with a particular property.

"Let's see what we did for Intermercado," he said. He opened the source code for Intermercado's store, looked at it, then entered the same property name into the new diagram.

Working the Web

I was trying to check my email when I realized that I had lost my network connection. I first thought it was just me, but the growing murmur throughout the room and the intensifying exchange of gazes confirmed that the network was down for everyone. The murmur soon transformed into a long collective *Iiiiiiii*—a humorous interjection somewhat similar to the English "oops," which then gave way to laugher and jokes. It was indeed a bit of an "oops" situation—a company predicated on using the newest software technology could hardly proceed without an Internet connection.

It was soon announced that power to the whole office would need to be turned on and off in order to restore the connection; everyone started shutting down computers. People began moving around, talking, joking. Many went to the common area, taking seats on the couch and in beanbags. We chatted about random things. The system administrator had already flipped the power switches and the Internet was back, but the people were still talking. "The Internet is back up, you know," said Felipe, the founder. His tone sounded half-jokingly apologetic: he seemed to realize that acting as an authority figure and calling on his employees to get back to work would sound funny in the midst of this free-spirited moment that seemed to spotlight Alta's startup culture. Indeed, the comment just drew laughter. Nobody hurried to get up.

I did not return to Alta next day, having scheduled some interviews related to Lua. When I arrived the day after, Fabio told me he had a new idea about what I could do. There was a new open source package called "Spring Web Flow," he explained, built on top of Spring, a Java framework that seemed to be on everyone's lips at the time. Fabio wanted me to try using Spring Web Flow to implement a shopping cart. I asked him if he had used it before. No, said Fabio. He hadn't. He wanted someone to try it and to build a proof of concept—a demo that showed that this was possible. And he thought that I could take this task. I was flattered that I was getting recruited into Alta's research and development efforts, but soon realized that Fabio likely picked the task because it was the most appropriate one for an unreliable worker like me. The new task was something I could do at my own pace. And if I were to give up on the task without completing it (as I eventually did), this would not impact the schedules for Alta's projects.

I got back to my laptop to try to get started on Spring Web Flow. In theory, this was easy: I just had to locate the framework on the web, download and install it, then find some tutorials on how to build a simple application. I was sure I would have no trouble finding documentation. After all, as many of my interviewees often said, the Internet had become "the world's greatest library" where one could find *anything*.

I soon realized, however, that before I could get anywhere, I would need to do a lot of basic setup. First I needed an appropriate web server, a piece of software that would do the hard work of capturing the requests for web pages coming from clients' browsers and translate those requests into terms that my application could understand more easily. I went back to Fabio, to ask him what Alta's developers used. "Use Jetty," Fabio told me, explaining that this was a new Java-compatible server that was much faster than the alternatives. I went back to my seat, googled "Jetty," and installed it following instructions I found on the web. I then installed Eclipse, a Java development tool I knew everyone at Alta was using. I realized that there had to be a way of starting Jetty from Eclipse, so I returned to Fabio's desk for further instructions. He told me to install Jetty Launcher from inside Eclipse. Seeing that I looked lost, another developer offered to show me. Here, he said, traversing Eclipse menus: you install it, then you go here, you put your Jetty Path here, then click here, then it runs.

I returned to my desk trying to reproduce what I saw. I did not get very far: my Jetty Launcher and my Eclipse did not seem to like each other. As I eventually understood, Jetty Launcher would not work with the most recent version of Jetty. I returned to Fabio several times with questions. At one point, another developer, "Leonardo," jumped in. Yes, he said, Jetty Launcher requires Jetty 5 and would not work with Jetty 6. I told him I saw a discussion of this on the web and that someone has offered a patch: a set of changes to Jetty Launcher that made it compatible with Jetty 6. I was hoping that Leonardo would tell me whether this method worked, but his response was disappointing. He had read about the patch but had not tried it; he was still using Jetty 5. He encouraged me to keep trying with Jetty 6, however, and to tell him if I managed to get it working. Someone had to be the first to use it, so it might as well be me.

I went back to my desk, spending more time reading documentation and forum posts, eventually deciding to give up on the latest version and go with Jetty 5, the same version everyone else was using. I could finally start Jetty from Eclipse, but I now needed to build an application. I tried sample applications from Jetty, Spring, and other projects, but none worked. There were too many moving parts and it was impossible to tell which was causing problems. As I headed out for the day, I stopped by Fabio's desk to discuss the matter briefly. Leonardo overheard us again. "Use struts-blank," he said. I was not quite sure what "struts-blank" was, but I figured a Google query would give me the answer.

I resumed my task the next morning. As I learned, "struts-blank" was a trivial application that would work on top of Struts, another Java framework. Using struts-blank resolved a round of problems, but introduced me to the next set. I spent more time reading what I could find on the web and approaching the issue from different angles. I was finding lots of relevant documents, though few were helpful. Spring documentation assumed a lot of knowledge that I did not have. Jetty had documentation for version 6, but not for version 5. Jetty Launcher was missing documentation all together. I returned to Fabio with questions several times, at one point asking him if he or anyone else at Alta had actually ever gotten all of those pieces working together. "No," he responded. "The whole point is to get it working."

As this episode illustrates, software work requires a peculiar fusion of globalized and localized activities. Much of that work involves interaction with software developed quite far away and documented in bits and pieces around the Internet. The software and the documentation are quite mobile. In theory, anyone with an Internet connection can download and use them. Downloading software never used before in the local context and getting it to work by following documentation on the Internet is part and parcel of software development. In my case, getting the components to work together was "the whole point," as Fabio pointed out. Fabio had heard of Spring Web Flow, likely by talking to other developers or by reading technology news. The system sounded promising, and could perhaps become important for Alta's future claims to be using the world's newest software technologies. Fabio did not at that point know what exactly it would take to actually make Spring Web Flow work. *That* was the task he was assigning to me.

A software developer who cannot use the Internet to find out how to solve a problem that is new to his colleagues is of little use to a company like Alta. A good developer would also be careful to not always pester colleagues with questions that can be answered with a web search even for things that the colleagues likely *do* know about. (My position in the company seemed to allow me to get away with a lot more questions than other developers could afford to ask.) On the other hand, in practice, the task of getting downloaded software to work benefits dramatically from proximity to people who have worked with it before. A single phrase uttered by a colleague can substitute for hours of Internet search and trial and error, stressed many of my interviewees. Software development consequently becomes an intensely local affair. Developers often seek balance between local and remote knowledge through an active exchange of what they call "pointers"—links and keywords that can be used to locate additional information online. Like Giddensian "symbolic tokens" (Giddens 1991) pointers become a tool of globalization, providing cross-references between the concrete local reality and the abstract world of online software and documentation. My quest started with one such pointer—the short phrase "Spring Web Flow," which led me to abundant (if somewhat unhelpful) documentation. My progress was later furthered by additional pointers I picked up along the way, such as Fabio's suggestion to use "Jetty" and Leonardo's suggestion to try "struts-blank."

The power of local advice and Internet documents both have much to do with the shared context of work. A few hours later, everything was *almost* working. I returned to Fabio and asked him what version of Java he was using. "I just do 'sudo aptitude install java,'" said Fabio. I returned to my laptop, and typed the four words at the Linux command prompt. A few seconds later I had the right version of Java. My simple project was finally running (though this was only a small step in the task that Fabio gave me). Fabio's four words "sudo aptitude install java" magically brought my laptop in sync with all of Alta's computers, making sure that the steps that had worked for Fabio and Leonardo would work for me as well. This synchronization was possible, however, because Alta's machines, configured in Brazil, and my laptop, configured in San Francisco, were already running essentially the same software, Ubuntu Linux 6.10. Continuing synchronization of practice was much simplified by the extent to which the context had been synchronized through earlier work, a long process the beginning of which I described in chapter 4.

A Local Affair

Around 6:00 p.m., Eduardo started gathering people. They would be having a cake, he explained to me. It was a company tradition: once a month they got a cake and congratulated all the people who had birthdays that month. This time it was just Leonardo, a recent PUC graduate who had been transitioning to management and was the most recent minority partner. Everyone gathered in the conference room. There were two trays of snacks, a cake, and several large bottles of Coke. Eduardo suggested that we should do introductions for new people. The first of the new people mentioned that he lived in Niteroi—a city across the bay from Rio, from where a large number of Alta's employees commuted. After that each of the new people was asked to say whether they thought Niteroi was a better city than Rio. When my turn came, I introduced myself but dodged the Rio–Niteroi question. "And what about Niteroi?" several people demanded. I gave a vague response, hinting at a preference for Rio. Fabio, a native of Niteroi, aimed a bottle cap at me. Amid loud demands for me to take a clear stand on this important issue, I ended my introduction even more vaguely, unwilling to step on the neighborhood sensibilities of Alta's global IT professionals, skipping the opportunity to act as a foreign judge of this local rivalry. "Vaseline," snickered Fabio as he put down the bottle cap.

Leonardo started cutting the cake. "Who are you going to give the first piece to?" asked several people. There was some suspense. "I will give the first two pieces at the same time," said Leonardo. He cut two pieces. "Those are for my team," he said, giving them to the two developers who worked under his supervision. He then cut a piece for Eduardo, who was slouching in a chair, looking over the team like a patriarch. He owed everything to Eduardo, explained Leonardo, exaggerating the tone and making a bit of joke out of his public acceptance of Eduardo's authority. The next piece went to Felipe, another cofounder. Luís, the third of the original partners, was not there, so there was again suspense as to who would be getting the next piece. It went to Fabio. The move caused a murmur. Eduardo and Felipe were unambiguously the bosses of the company. Fabio and Leonardo, on the other hand, were both recent minority partners. Leonardo's move thus appeared to acknowledge Fabio's status, while also highlighting the difference between minority partners and everyone else. Leonardo laughed as he gave Fabio his piece, then put down the knife: others could cut their own pieces. Startup spirit aside, Alta did have founders, minority partners, and general employees. Fabio and Leonardo had to learn to manage their new status vis-à-vis others.

During one of my last weeks in Rio in 2007, I met with Rodrigo Miranda at a café in Copacabana. He had agreed to provide his comments on a paper I was going to present at a conference upon my return to the United States. One of the things he mentioned concerned my discussion of how developer build ties to the remote centers of software practice. Rodrigo suggested that "build ties" was perhaps too strong a phrase. Most people adopted foreign technology and got quite good at it, explained Rodrigo. But this had little to do with building actual social ties to foreign communities, he continued. He had been trying to do it with Kepler and finding it extremely difficult. Most people never tried. Look at Alta, he said. They adopted the culture, but without the social ties. Their clients are and may always be local.

Rodrigo was right to some extent. Alta was an intensely local affair. The company used foreign technology and lot of foreign culture, without much direct contact with the foreign centers. In contrast to Kepler and Fabio's Lua projects, carried out on the side for "fun," Alta's main line of work was local, both in its location (involving little interaction with people outside Brazil) and in its significance. Such local work was often boring and brought the developers limited cultural dividends in the larger world of software practice. In many ways it came down to sales engineering: doing what had to be done locally to allow a Brazilian company like Intermercado to use the software supplied by EIT.

The local focus of Alta's work, however, was also a source of strength. In addition to making a profit, Alta was building IT solutions that were actually being used by many people. After my departure in 2007, the company proceeded to strengthen its relationship with Intermercado, eventually winning the bid to write software that would control the front page of one the most popular online stores in Brazil. The months that followed were a period of much work and much learning, Fabio told me when I returned in 2008. He and other Alta engineers had to learn to build a web application that could handle traffic never faced by any of their earlier applications—or, quite possibly, by any software based on Kepler, I can add. Another local contractor of Intermercado, required by Intermercado to work together with Fabio's team, was instrumental in this learning. Local focus was bringing Alta to projects whose scale made them exciting and a great source of war stories. In the same months, Fabio finally stopped participating in Lua-related projects. There just was no time for such games.

6 Porting Lua

In 1993 a group of computer scientists working at a university in Rio de Janeiro developed a simple programming language called "Lua" to serve the needs of a Brazilian company based in the same city. Nineteen years later, Lua is often ranked among twenty of the world's most popular programming languages[1] (out of thousands) and has a user community spanning five continents. While Lua has brought its authors rather modest financial rewards (it is distributed for free and brings little consulting revenue), its use in popular software such as Adobe Lightroom, *World of Warcraft*, and, more recently, *Angry Birds*, has made it in some ways one of the most successful software products ever developed in Latin America. Lua's story provides us with a rather different picture of peripheral participation in a global world of practice than the case of Alta that I discussed in chapter 5. This picture is also a lot less intuitive and more complex. I therefore look at Lua extensively in two chapters: this one and the one that follows.

One of the things that makes Lua's story unintuitive is the fact that the language is little used in Brazil. In 2007, when I was doing my fieldwork, few Rio programmers had heard of it. The situation has changed only somewhat since—Lua is now better known, but still rarely used. This isolation from the local context, however, is the flip side of Lua's success. American users of Lua often credit it with being highly *portable*—Lua can run on many different computing platforms. While increased portability in this narrow technical sense is an important part of Lua's story, I focus here on a different kind of "portability": Lua's gradual transformation from a highly local project to an international programming language that betrays little connection to the city where it was developed and where it is still based. I organize my discussion around Giddens's (1991) notion of "disembedding"—the "lifting out" of social (or in our case socio-technical) relations from their local context, which then makes them mobile across time and space. Following Lua's transition from a highly embedded project—developed as a

solution for a specific set of problems, entangled in a web of local relations, goals, and commitments, and reliant on "tacit knowledge" (Polanyi
1966; MacKenzie and Spinardi 1995)—to an international programming
language, we observe the different mechanisms that enabled and facilitated
this disembedding.

As we will see, Lua's disembedding and its later international success
were not planned in advance. To a large extent the disembedding of Lua
simply "happened," in many ways without a conscious intention by its
authors. It happened in part due to numerous decisions that most participants saw as quite natural. In some of the cases, acting otherwise—for
example, using Portuguese words as Lua's keywords—would be nothing
short of silly according to some of my interviewees. It is important, however, to look closely at such "obvious" decisions. It is by understanding
how such decisions come to be obvious, and why they are obvious to some
and not others, that we can come to see the geographic logic woven into
the professional culture of software development.

The story of disembedding told in this chapter complements the investigation of local reassembly of a foreign practice presented in chapters 4 and
5. After decades of work that helped establish the foundation of software
practice in Brazil, the context was created that made it possible for some
of the practitioners to engage in one of the most central roles in the world
of software: developing a new programming language. This replicated context, however, is characterized by a distinct pattern of connections that
makes it different from the remote original in many ways. Brazilian academic computer science has strong connections to foreign computer science, which proved an important factor in Lua's success. At the same time,
much *unlike* the American computer science research community, which
is famous for tight linkages with industry, Brazilian academic computer
science is relatively isolated from both local and foreign computer industry
and instead exists in somewhat of an enclave. This makes the experience of
Lua's authors quite different from that of their students working for companies like Alta, whose success depends in many ways on tight integration
with local systems of production.

I start this chapter with a look at my interviews with Lua's users in California in 2007. I then return to Lua's history from the early 1990s, proceeding to around 2003, a point at which Lua could be said to have achieved
its fullest disembedding and was starting to become a major success. In the
next chapter, I then turn to the limitations of this process of disembedding,
looking at Lua's changing relationship with the university, city, and country where it was born.

Choosing Lua

"Craig," one of the people who responded to my request for interviews sent to the Lua mailing list in early 2007, was an engineer at a small startup in California. Like most other users of Lua whom I interviewed in California, Craig encountered Lua online, while searching for a scripting language to embed in the software application he was working on, an online computer game. None of the people he knew personally at that point had heard of Lua before. Craig mentioned Lua's small size and simplicity as the reasons for choosing it over a "more mature" language such as Python. He was particularly concerned with the security of his application and felt that a smaller and simpler language, one that he could understand more thoroughly, would help. Craig noted an important weakness of Lua—the relative scarcity of libraries, a hurdle faced by all new languages. This factor, however, was of little relevance to him. "We were not building an application [in Lua], like a web server or something else that would need a whole bunch of specialized libraries," he said. They just needed a way to add scripting to an application they already had, written in C++.

Craig's use of Lua was quite typical: most Lua users in the United States employ Lua for "scripting" applications written in C, a programming language developed in the early 1970s, which came to dominate software development by the early 1990s. Over the last decade and a half, however, many developers have moved to newer languages such as Java, Python, or more recently Ruby. Those newer languages use a technology that relies on what is called a "virtual machine" (or VM)—a software layer that provides a degree of isolation between the programmer's code and the machine's hardware. This isolation makes it possible to develop software much more quickly, though at a price: the resulting software runs more slowly. Consequently, C and its close relative C++ remain popular, especially for the kind of software where speed is important.[2]

In theory, most of the new VM-based languages can be combined with C in a single application, potentially allowing the developer to get the best of both worlds. In such a hybrid design, some parts of the system would be written in C, while other parts would be "scripted" in some language that allows for easier development. (The words "scripting" and "programming" mostly mean the same thing, except that scripting usually suggests easier work that does not delve as deeply into the innards of the machine.) Such usage is often complicated and is frequently frowned upon. For example, Sun Microsystems had pursued a targeted campaign to eradicate such mixed applications involving Java, encouraging the programmers to write their code in "100 percent pure Java." Lua, on the other hand, generally presents itself as the language primarily designed to be used together with C.

Lua has become particularly popular in the development of computer games, where efficient use of computer hardware is often crucial and the developers frequently work closely with C modules for handling graphics. Lua allows such developers to use "easy" Lua for parts of the code that are likely to change often, while relying on the more efficient C for tasks that are most likely to put strain on a computer's resources. Lua thus thrives in a relatively small niche, where it has positioned itself as a complement to a well-accepted technology, offering certain unique features that shield it from devastating competition with the "more mature" languages such as Python. Developers like Craig note and appreciate Lua's suitability for computer game development.

Like other interviewees, Craig found it quite easy to get started with Lua. My question about how he learned to use Lua and what kind of resources he used took him by surprise. "I am sure I downloaded everything, ran the command line, found out how that works," he said after a pause. Another interviewee "Steve," a lead engineer for a team of software developers working for a large software company in California, reported similar ease, which he then contrasted with JavaScript, a programming language developed by Netscape that my interviewees often considered an alternative to Lua. He did not need to look for people who understood Lua, Steve told me. Had he decided to use JavaScript instead, Steve would have gone and talked to people in his company who had a JavaScript implementation. "But that's because JavaScript is messy," he explained. "The great thing about Lua is that you don't need any of that." Lua's elegant simplicity made its foreign origin irrelevant—Steve and Craig could use Lua even if nobody else in California did so.

Like Steve, Craig did not feel constrained by his lack of contact with other Lua users. Just as he was starting to work with Lua in the summer of 2005, however, a Lua workshop was organized at the Adobe office in San Jose, about twenty miles south of where Craig's startup is located. (Adobe itself was extensively using Lua in one of its projects—Adobe Photoshop Lightroom—which was released two years later.) The event included presentations by two of Lua's authors, and Craig attended a part of it. "My reason to go was to get some sense of how serious this is," he explained. "To ask a few questions. To talk with some people about it." Seeing live users of Lua helped Craig feel more confident about his choice. He was not, however, looking to build ties with those people, he explained. At least as far as any technical questions regarding Lua were concerned, Lua's documentation and online community provided Craig with all he needed.

A Global Perspective

All American users of Lua I talked to said that Lua's Brazilian origin was nearly irrelevant to them. "I take a global perspective on those things," said "Rich," a software contractor who used Lua in his projects whenever he had a chance. Taking a global perspective on Lua luckily took little effort. While none of my California interviewees could read Portuguese, this presented no problem when it came to using a language developed in Brazil, since the Lua community interacted in English. "I guess English language is the lingua franca for Lua as well, from what I can tell," explained Craig. "I haven't seen any Brazilian, or Portuguese, emails coming up." Therefore, he explained, he had never had any concerns about being able to access help or documentation.

Emails in Portuguese *do* occasionally arrive through the Lua mailing list, and are typically treated politely, usually receiving a reply in English (sometimes quoting the original question run through an online translator) and occasionally even in Portuguese. Displaying a global perspective simply requires treating such infrequent occurrences with humor. The discussion on the Lua list shows that some of the list members have a certain curiosity about Lua's unusual origin and sometimes allow themselves friendly questions about Brazilian practices that they find surprising—for example, the frequent use of a pair of brackets to close email messages (as a shorthand for "hugs"). Those who want to, however, can simply ignore such cultural differences. Most of the translation work is already done by the Brazilian members of the list.

Perhaps the only way that Lua's foreign base factored into Craig's decision making was because of the increased difficulty of understanding how "academic" Lua was. Craig knew that Lua was produced by university researchers and was worried that it was "an academic exercise." Making this evaluation would have been somewhat easier in the case of an American university—for example, nearby Stanford—Craig explained, since he could meet the authors or would perhaps "know somebody who knows somebody who knows them." In that sense, Lua was more "opaque." However, Craig ultimately found other ways to understand the intentions of Lua's authors: reading about the language, using it, later attending the workshop, and, perhaps most important, learning that Lua had been used in *World of Warcraft*, one of the most successful computer games at the time.

Other interviewees similarly expressed the desire to understand the people behind Lua as a way of learning where the language may be going in future, and some noted that not being able to *see* the authors (at least on video) made it harder to make this judgment. For many, however, Roberto Ierusalimschy's book provided enough answers:

Rich: Like Perl, the book was written by the architect of the language. *Programming in Perl* is written by Larry Wall. And the Lua book was written by Roberto. So, by reading the book you get the sense of both the designer's personality and the language itself. And so just reading the book it was pretty clear that the guy who made this language was a really smart guy, and he valued principles that I valued: simplicity, elegance. And at that point it doesn't really matter where he lives or what nationality he is. He is just a smart guy who made a good language, that's all that matters to me. If he wasn't able to speak English, it probably would have been a problem. But obviously there wasn't any communication barrier. So the fact that it was made in Brazil wasn't anything to me.

Lua authors' ability to successfully communicate (in English) their commitment to the principles that potential users shared made Lua's geographic origin irrelevant as far as most foreign users were concerned.

American users of Lua also found that they could understand Lua by placing it in the larger genealogy of programming languages, most of which were developed either in the United States or with the strong involvement of American computer scientists. Steve, for example, described Lua as "Lisp-like," linking it to a venerable programming language developed at MIT in the 1960s. Other developers described Lua by producing a list of its features. "Suffice it to say," said one 2007 article on Lua, "that Lua is an elegant, easy-to-learn language with a mostly procedural syntax, featuring automatic memory management, full lexical scoping, closures, iterators, coroutines, proper tail calls, and extremely practical data-handling using associative arrays" (Hirschi 2007). Such a description again helps make Lua's geographic origin irrelevant by placing the language firmly within a classification system developed primarily by American computer scientists.

The classification system that makes it possible to describe Lua using a list of eight concepts, as Hirschi does in the passage quoted previously, and the larger system of meaning to which it is linked provide an important disembedding mechanism. A programming language that is designed in such a way that it can be explained in terms of this shared system is relatively free to travel. Such academic sophistication is not expected of all programming languages—some of them are notorious for being "messy" and "ugly" and become widespread primarily through a strong association with a powerful actor. For Lua, however, its academic credentials were crucial.

Users of Lua often also point to another important disembedding mechanism: Lua's internationalized credentials typically expressed in the form "Lua is used by X," where X would be a major company, or "Y was written in Lua," where Y would be a well-known product. Sometimes, developers cite such credentials as being important for their own decision to go with Lua. Others draw on such credentials to convince their colleagues. Steve, for example, told me that his choice of Lua of course attracted questions within the company. "What is this? Where does this come from?" asked his colleagues. Steve explained that Microsoft had shipped products with Lua.

"Used by Microsoft" can be understood as an instance of what Giddens (1991) calls "symbolic tokens." Symbolic tokens represent known units of value recognized across space and act as an important disembedding mechanism, by removing the need for situated trust. "Used by Microsoft" deflects the original question about Lua's origin ("Where does this come from?") and whether its authors can be trusted. If major companies use Lua, where it comes from and who wrote it are a lot less important.

Legacy Stuff

Over its history, Lua has undergone some substantial modifications, often making old code incompatible with the newer versions of Lua. Craig remembered this issue being discussed at the workshop he attended. "During the meeting a lot of people were worrying about legacy stuff," he said. As a new user, however, he had little interest in the topic. "And in my case, I was just like: 'Oh, I don't care, please break compatibility, make it good for *me*!'" he explained.

Existence of old code ("legacy stuff") creates a serious problem in software development. When programming languages and software libraries are put to use, their limitations eventually become clear. While those limitations can often be overcome with additional code, such incremental additions lead to increasingly "ugly" and "ad hoc" design. The authors face the temptation of rethinking their design and "cleaning it up." To be truly effective, such "cleaning up" often requires radical changes, which would make the new version incompatible with code written earlier. Changing old code to work with the new version of the programming language or a library often requires a lot of work and introduces a new opportunity for bugs. It is therefore not taken lightly. Avoiding the need for such changes is called "backward compatibility." The need for backward compatibility typically impedes the evolution of a language or a library and leads to an increase in the size of the code ("bloat") and unnecessary complexity ("cruft").

While other Lua users in California expressed somewhat more interest in backward compatibility than Craig, most considered Lua's willingness to break with the past to be a major source of strength, comparing Lua explicitly with JavaScript, a language embedded in most modern web browsers. Lua devotees sometimes described JavaScript as "nasty," comparing it to Lua's "elegance." Some pointed out, however, that this "nastiness" was very much connected with JavaScript's widespread adoption:

Rich: JavaScript suffered from premature standardization. There is this web browser, the web blew up, so everybody is using JavaScript. And then they thought: "Oh, we have to standardize this, make it interoperable." And the language hadn't really stabilized at that point. And so there is a lot of cruft and nasty things. Both in JavaScript, that is the language itself, such as the "with" operator, and especially in the object libraries. [. . .] *Because* it was standardized so early, *because* it had a huge community. Whereas Lua didn't really have those forces. It had a small community and less momentum. So the designers could, when they realized they made a mistake, throw it out, unify the concepts under a different way of thinking. The different abstractions that went from Lua 3 to 4 to 5. So, JavaScript *could* have been as good a language as Lua, I think, if it hadn't had this pressure on it, by the huge community, a huge user base.

For Rich, Lua was simple and elegant because it was not weighed down by commitments to any existing body of code. Not having such commitments was, however, a choice that Lua's designers had made, as they broke compatibility with Lua's original applications. Comparing Lua and JavaScript, we should consider that while popular web browsers had prevented JavaScript from achieving Lua's elegance, it was those same browsers that made JavaScript relevant. Cutting the links with those browsers would undermine JavaScript's popularity. Lua's early commitments, on the other hand, were to custom software written for a company located in Rio de Janeiro. Most users of Lua agree that leaving such a "legacy" behind was crucial for Lua's international success. As we will see, some users in Rio de Janeiro had to bear the cost.

DEL, SOL, LUA

If we went back in time to the early 1990s, when the first version of Lua was created, we would find a highly embedded project, one tied to local goals, relationships, and commitments. Had Craig tried to use an early version of Lua, he would have likely found it a daunting task. He would have faced a piece of code written for a specific purpose (quite unlike his), offering him little in terms of social support and no justification if his choice of the language were questioned. At the same time, as we will see, from the earliest days of Lua, the authors had made a number of choices that allowed for future disembedding.

In 1992, Roberto Ierusalimschy, a native Carioca, returned to Rio de Janeiro from a one-year postdoc at the University of Waterloo, in Canada, and started working as an assistant professor at PUC-Rio's Department of Informatics, where he had completed his PhD two years earlier. In addition to his job as a professor and his academic research in programming languages (working on an experimental language that "completely failed," according to him), Roberto started doing consulting work at a PUC-based consulting venture called Tecgraf.[3] In the early 1990s, Tecgraf was a fairly small group of PUC students and professors offering IT consulting services to a number of organizations, including Petrobras, Brazil's main oil company and one the country's largest corporations. Petrobras was an unusual client—a semi-public company responsible for reducing Brazil's dependence on foreign oil by developing the capacity to extract deep-sea oil off the coast of Brazil (and especially in the areas surrounding Rio de Janeiro). Petrobras thus faced substantial technological challenges and was an important consumer of scientific expertise.

Also at Tecgraf was Luiz Henrique de Figueiredo who had also just recently received his PhD in Rio de Janeiro at the Institute for Pure and Applied Mathematics (IMPA) and was employed full-time by Tecgraf. Also a native of Rio and a graduate of PUC-Rio, Luiz Henrique had earlier spent three years pursuing a PhD in England and another year working at the University of Waterloo. Luiz Henrique was trained as a mathematician, thus benefiting from a different and considerably longer history of efforts to transplant a scientific practice into Brazil. From his early days as an undergraduate at PUC, however, Luiz Henrique had been interested in computing. After returning from England, Luiz Henrique later spent a year in Canada, working at the University of Waterloo's Computer Systems Group. He focused his doctoral research on computer graphics, while working as a software developer at Tecgraf.

In 1992, Luiz Henrique turned to the problem of providing a unified way of configuring graphic interfaces for a large number of software applications that Petrobras used for simulations related to oil extraction. "These were huge programs that were very old and very refined, and they didn't want to give them up," said Luiz Henrique, "But at the same time, because they were very old, the interface was very clunky." Tecgraf was asked to provide a better interface for this old simulation software, something that would allow the users to simply click on a diagram, enter a value, and request a simulation. Realizing that Tecgraf would need to provide such an interface for a wide number of simulators, Luiz Henrique started thinking about developing a language for expressing the configurations. "This was kind of a typical problem," he explained to me. "You would write a simple text file that would say: I want this diagram and in this diagram when I click this entity you should show this kind of a menu and do this kind of data validation, things like that. And then when I am done I want this data to be output in this format." The language was developed in 1992 and was called "DEL," short for data-entry language. It was what would today be known as "a domain-specific language"—that is, a language intended for a highly specific purpose, in this case configuring oil extraction simulations.

While DEL was a success in Tecgraf and among its users in Petrobras, it soon became clear that it was too limited to build all the applications that Petrobras wanted. In mid-1993 Luiz Henrique met with Roberto and Waldemar Celes, then a PhD student at PUC, who had themselves developed another domain-specific language called SOL (Simple Object Language) for another of Petrobras's many specific problems, which had also been found too limited. The outcome of the meeting was a decision to replace both DEL and SOL with a new language, which was soon implemented by Waldemar as a course project. The new language was called LUA, meaning "moon" in Portuguese—a pun on SOL (Portuguese for "sun"), but also, as somewhat of a joke, an abbreviation for Portuguese "Linguagem para Usuarios de Aplicação"—"Language for Application Users." LUA was a success and was quickly picked up by other projects at Tecgraf.

Comments in English

DEL, SOL, and LUA (soon renamed "Lua") were all written in the C programming language, which was at the time and still is the lingua franca of programming languages. Similarly, from early on, Lua displayed a commitment to another lingua franca: English. Lua uses English keywords as its basic vocabulary. Its code is also written in English, which includes the names of variables as well as the comments.[4] Lua's documentation is also provided primarily in English.

Lua's authors had somewhat different memories of the decision process that led to their choice of English over Portuguese. As described in chapter 2, Roberto recalled a discussion that weighed pros and cons, with the eventual choice being driven by the practicality of diacritic-free English, a preference for "standard" keywords, and the desire for consistency when it came to comments and error messages (i.e., those were in English to match the English keywords). Luiz Henrique, by contrast, did not remember the topic ever being the subject of discussion, seeing use of English as the only sensible option. "No one would take Lua seriously if it had Portuguese keywords," he explained.

To get a better understanding of the logic that might have led to the choice of English, I have also talked about this with some of Lua's early users. One of them, a Tecgraf developer who worked with Lua "when it was still called SOL," first told me Lua-like use of English has always been a standard practice at Tecgraf. He then corrected himself, however:

Antônio: No, sorry, the *comments* are in Portuguese . . . obviously, right? But variable names, we usually try to do all in English. But our comments are in Portuguese. Lua is different.

As Antônio then explained, using Portuguese for comments would be undesirable, as it would potentially show the limitations of developers' English skills. "We don't even *want* people here to try to write comments in English, because most of them are not fluent in English, so it would end up being broken English," he said. Unlike code, which uses English in a constrained way, with heavy reliance on abbreviations (e.g., "cal" or "apl_unsel_group"), comments are written in full sentences. They consequently offer little opportunity to hide any weaknesses in a programmer's English. Using English for variable names and providing comments in Portuguese provides a sensible (and very common) compromise. (Antônio's explanation of why the code itself should be in English were similar to those discussed in chapter 2.)

Lua was different. The first distinction, not stated but implied in Antônio's analysis, involved the developers' proficiency in English. Unlike the Tecgraf programmers who, Antônio feared, might write comments in broken English, Luiz Henrique and Roberto were both comfortable with English, as well with the English-speaking academic culture more broadly, having spent some time abroad. Lua's authors could thus write their comments in English without the risk of embarrassing themselves with simple mistakes. This competence (and confidence) with English gave Lua an additional early start on disembedding.

The second distinction concerned Lua's need for mobility.

Antônio: In Lua's case I think it even makes sense for everything to be in English. It was born to fly, for other. . . It has a much more globalized use than our applications.

He contrasted this to Tecgraf's other software:

Antônio: Our products are not . . . Our source code is not for export, it's not open. What we do is not open. The code we write is for Petrobras and is their property, and they don't want to have to also . . . to know English to read our code. So, for several reasons, in applications we don't write comments in English. Variable names—yes, in the code. You read it more or less in English. But comments . . .

Unlike other applications, Lua was "born to fly," as Antônio saw it. Consequently, he did not remember anyone ever finding its use of English strange.

When I pointed out to Antônio that Lua was originally clearly *not* meant to "fly," but was ostensibly written to solve a specific problem faced by Tecgraf, he offered a different explanation of the Lua team's possible rationale by referring to his own project, which he saw as *potentially* open.

Antônio: Yes, nobody could think about . . . the explosion, the success that Lua would have. The acceptance in the *games* [says in English] industry. But maybe they, Roberto and Luiz Henrique, had this idea, I don't know. For example, this [hobby project] that I wrote, that's all in English. Comments are in English. Because in my case I was thinking, I don't know: one day I'll put it on LuaForge, someone will want to download it. I like this idea of *open source* [says in English], of *many eyes* [says in English].[5] Everyone will be looking. I think that when you do an open source application, you have to speak the most widespread language, the most common language, most easily understood. But not for our applications . . . Not at Tecgraf. We actually don't want this to happen.

While Antônio said that he never wrote comments in English when working on Tecgraf products, he used English comments in his hobby project. He made it clear that he neither expected this project to become anything big, and in fact he had not even gotten around to releasing the code to anyone. "One day," however, he could put it on LuaForge, a web site where users of Lua share code. And who knew, perhaps it would become popular. Using English kept open this possibility.

Antônio's project and Lua were both different from the products that Tecgraf built for Petrobras by their relative isolation from the local power relations. Neither was written to immediately become property of a company. While both projects perhaps technically belonged to PUC, as an academic institution PUC seemed content to let the developers treat their code as free. (See the next chapter, though, on the question of PUC's ownership of Lua.) In addition to simply giving them the freedom to write such projects the way they wanted to write them, this made the possibility of a global success somewhat more imaginable, since the projects' destinies were not as obviously in the hands of the bureaucracy of a Brazilian corporation

To understand Lua's early use of English, we must thus consider the *potentiality* of Lua's later success, the inherent ambivalence of such projects, and the notion of "subvocal imagination" that I mentioned in chapter 1. Created for specific needs (often the needs of those who pay the bills), projects such as Lua may from the beginning carry the imaginable *possibility* of global success. While such a global future is rarely *planned* for explicitly, it can be *imagined*, and this may be sufficient reason for developing the project in a way that would not altogether preclude the possibility of a global success. (I return to this topic in chapter 8.) This means, among other things, writing comments in English.

When explaining his rationale for writing his hobby project in English, Antônio drew explicitly on *open source* terminology. Today, the open source paradigm provides developers with a ready vocabulary and an accepted framework for explaining such decisions. While Lua eventually became "free software," it did not *start* this way nor does it appear that the free software / open source vocabulary was available to Lua's authors in 1993. I look at Lua's transformation into a free software project in the next section and come back to this issue in the next chapter.

Let's See What Happens

Lua's authors stressed in our interviews that the language was developed to solve specific problems Tecgraf faced in its work for Petrobras, and that Lua's later international success came as a major surprise to them. When I asked whether his personal goals for Lua had changed over time, he responded as follows:

Roberto: Completely, completely. Completely! This is so huge, I can't . . . It changes everything. When we started Lua . . . This is one of the things that people do not . . . When we say in our paper, that paper about the history of Lua,[6] that it went beyond our most optimistic expectations, this is not very true. Because we didn't have *any* expectations. [. . .] We really created a language to solve this specific problem we had at the time. That's why we joke, but it's true: There was no "Lua 1.0."[7] There was "Lua." We did code, and that worked and "oh great, it solved our problems here."

Lua represented a specific solution for a specific problem, and Lua's authors did not attempt to innovate for the sake of innovating. Lua offered a pragmatic combination of features, informed by academic research on programming language design, but not aiming to contribute to it.[8]

The pressure to publish, however, combined with Lua's early success at Tecgraf, soon led the authors to showcase the language outside Tecgraf,[9] starting with a short presentation at a conference in Brazil in 1993. "We are academics," explained Roberto. So, when they heard of a conference where they could present their work on Lua, it seemed like a natural decision, especially since the conference had an event that aimed to include "real" applications, and Lua, which at the time was used by eight to ten real people at Petrobras, was exceptionally "real" by the standards of academic research. According to Roberto, the presentation was extremely well received.

While Lua was not written to advance computer science research, the authors' academic backgrounds helped them *explain* Lua in terms of current computer science research, making it possible for Lua to travel away from Rio using a set of academic papers as a vehicle. A few years later the same ability to explain Lua in proper terms became important in giving Lua credibility in the eyes of American software developers working outside academia.

In 1994, the team wrote a longer paper for another Brazilian conference. This time, the paper was written in English and referred to the language as "Lua" rather than "LUA,"[10] avoiding the hassle of explaining a Portuguese acronym in an English paper. The paper also included a link to download Lua ("Lua 1.1"). A colleague at Tecgraf "pushed" them to put Lua up on a web site and to include a link in the paper "to show that this wasn't just vaporware," Roberto explained. ("Vaporware" is a developers' term for software that is described but does not actually exist or does not work as described.) Encouraged by Lua's success so far, the team was also curious to see what might happen next, and Luiz Henrique announced Lua on a number of newsgroups.[11] "And then some people started using it," said Roberto. "But it was kind of, 'Let's see what happens.'" The paper and the announcement started a slow trickle of questions, some of them from abroad.

Lua 1.1 was packaged with an informal license that allowed free academic use but reserved the rights for commercial use:

Roberto: Something we wanted, that I remember . . . Again someone gave us this idea to try to sell Lua. In the beginning we put it on the Internet with a free academic license, and "Please contact us for commercial use." So there was this idea "Let's try to sell Lua." And then it stayed this way one year and we got one contact. [Laughs.] Without success. Just a contact for "Maybe we could use . . ." for commercial use. So, we decided we were not going to sell it. But after that we noticed that there were people using it and people were liking it, and we were liking that idea of other people using Lua.

Lua 2.1, released in the early 1995, included a license written in proper English "legalese" and allowing almost unrestricted use of Lua for both academic and commercial purposes.[12]

The change of license not only gave additional freedom to Lua's potential users, but also signified the authors' changing perspective on what they could and could not achieve with Lua. Having started with a rather vague idea of what Lua could lead to and originally holding open the possibility of selling it for money, they moved to "liking that idea of other people using Lua." He then continued:

Roberto: That kind of . . . touched . . . satisfying, a kind of gratification for us, gratifying, whatever. And so we started to feel good about that. [. . .] And then we published the article: "Let see, let's try to get more users, to promote Lua." So we put up . . . And then the reaction was very strong, and then it started to be really important—the outside users.

Consistent with Becker's theory of motivation discussed in chapter 2, Lua's authors did not start off with the intention of distributing free software, but rather developed the appropriate "perceptions and judgments" as they engaged in the activity. As members of an academic community, however, and in particular being fairly fluent in the culture of Anglo-American computer science, from which the free software movement got much of its culture, they were of course well prepared for developing such perceptions and judgments.

In 1996, the team published an article about Lua in an American computer science journal (Ierusalimschy, Figueiredo, and Celes 1996), as well as another one in a popular magazine widely read by American software professionals (Figueiredo, Ierusalimschy, and Celes 1996). Steve, one of my California interviewees, remembers originally learning about Lua from one of those two articles.

The two 1996 publications resulted in an increased stream of questions and the decision to set up a mailing list.

Luiz Henrique: Around that time, I remember now, we wrote this article in *Dr. Dobb's Journal* and from then on we started to get messages from abroad, people asking questions about Lua. So, we thought, well, maybe we are going to get too many questions and won't have time to answer them all. So we created the mailing list for that, so that other people could answer our questions. [. . .] Maybe Lua is going to get some interest, and how about creating a community? [. . .] If we were going to get a community, maybe we should have a mailing list so that they could talk among themselves? To not have to answer everyone individually.

The list (*lua-l*) was set up in February 1997.[13] With the addition of *lua-l*, the language was increasingly starting to look more like a free software project.

Prior to the creation of *lua-l*, the project had no dedicated mailing list. (The early users remember no need for it, pointing out that all people using Lua were by and large in the same place, and often in the same room.) The English *lua-l* thus quickly became the central forum for the Lua community, attracting many Lua users from PUC, as well as a smaller number from other Brazilian research institutions. In 1997 the Brazilians (people with .br email addresses) comprised a little under a quarter of the list's participants, constituting the list's largest "minority." As the list grew, however, the percentage of Brazilian participants started to decline, eventually getting overtaken by Germans in 2007.

More Exciting Users

A month before the mailing list was set up, Lua's authors received a message from a programmer working for LucasArts, who wanted to congratulate them on developing Lua ("Its elegance and simplicity astound me," he wrote) and to let them know that he was thinking of using Lua in one of LucasArts' games.[14] Once the list was created, the programmer became an active participant and in April revealed to the list that he was working on a "scripted adventure game engine." Soon, other users started to discuss their use of Lua for scripting computer games. A year later, LucasArts released a game called *Grim Fandango*, which became Lua's first international success case—not quite "used by Microsoft" but a major step toward it.

Grim Fandango also gave it a new "place." Lua had a new, international "origin," now associated with a community that was not narrowly localized. *People ask where Lua is from*, said Steve. *But they do not usually mean location, but rather which industry or context. For example, JavaScript comes out of the Web. So, I give two answers: "PUC-Rio" and "the games industry."* While the games industry was not really the Lua's origin, he then explained, it provided a context in which Lua can be understood. *People don't care who wrote it*, he said. *They want to know how it fits into the world.* The games industry thus provided Lua with a place where it could belong in a foreign context.

Lua's success at LucasArts also gave Lua new advocates, located in about the best place for promoting Lua. As the LucasArts programmer explained on the Lua mailing list in 2001, face-to-face interactions in California were instrumental in helping Lua gain popularity in the games industry. In 1998, some of the LucasArts programmers attended the Game Developer's Conference, the largest trade event for computer games developers, held annually in the San Francisco Bay Area. One of them made a presentation about implementing scripting languages for computer games. The talk was delivered to an audience of two hundred to three hundred people and focused on the difficulty of developing a good scripting language. In the end, though, the speaker discussed LucasArts' use of Lua. "People lit up and furiously started scribbling notes and looked really excited. I got a few inquiries afterwards, but game developers being who they are, most of them just went out and checked it out on their own. Soon enough the list was overflowing with game programmer inquiries."[15] A presentation delivered at a conference in the very center of the software world brought Lua to the attention of the larger gaming industry. This attention eventually led to the use of Lua in a game released by Microsoft Games, giving Lua the crucial "used by Microsoft" status.

The growing list increasingly became an important source of influence on Lua. "Tecgraf was kind of stable and was not demanding that much," said Roberto. While having a nondemanding employer may be a blessing in many lines of work, this is not necessarily the case in software development. Finding that Tecgraf was no longer presenting serious changes, the Lua team increasingly turned its attention to the outside. "So I think it was also kind of like: 'Let's try to find more exciting users,'" said Roberto. The list members supplied the desired challenge, by applying Lua to new domains and running it on new platforms. Lua's gradual "porting" to foreign contexts went hand in hand with porting in the more narrow technical sense: Lua was increasingly used on computing platforms that were never used by Tecgraf, from tiny computing devices to the Cray supercomputer.

Roberto: That put more pressure to make Lua really portable. In the beginning our goal of portability was Tecgraf's set of computers. So, that was our goal, must run on that. It was a very large variety of computers that Tecgraf had, so from the beginning it was very portable. So it must run on DEC, on VAX, on *ta ta ta* [etc.]. And then later when people . . . I remember in '98 someone wrote and said they ported Lua to Cray, the supercomputer Cray 1. That was very exciting: "Wow, Lua is running on Cray." And then these things started, for instance, to show us that we must really think about ANSI C and about real standards. And not about "It runs on those machines and that's good enough." For instance, this was something that came from outside.

Paradoxically, Lua's origin in Brazil offered it an early start on portability. Due to the restrictions imposed by the market reserve, Petrobras had a limited choice of what computers it used, having to buy from different manufacturers depending on who had been allowed to bring computers into Brazil in a particular year. Over the years, it had accumulated a rather diverse collection of computers, requiring that Lua be run on all of them (Ierusalimschy, Figueiredo, and Celes 2007). This collection did not include things like the Cray supercomputer, however.

Demanding and attentive users are often considered in open source communities to be a valuable resource per se, as they help to "push" the project forward, providing feedback and gratification to the authors. In most successful projects users also contribute code, essentially becoming codevelopers. Despite releasing Lua under a free software license, Lua's authors never embraced the community-driven approach to software development—all modifications to Lua's code have always been done by the three members of the Lua team. (I discuss the reasons in the next chapter.) Lua's users, however, contributed to Lua in many other ways. Many became active members of the mailing list, helping answer questions and contributing ideas and resources. In 2000, one of the list members organized a wiki, which became an important resource for the community. (Steve cited the wiki as the source of his knowledge that Lua had been used by Microsoft.) A year later, another member of the list purchased the domain name "lua.org" and donated it to the team. (Since 2004 another user has volunteered to host the lua.org web site on his company's servers in the United Kingdom.) The list members also offered substantial help with the first edition of Roberto's *Programming in Lua*. Such contributions helped further establish Lua in its new position: a highly portable programming language easily embeddable in C applications such as computer games, supported by a networked community.

Breaking from Tecgraf

During the first decade of its existence, Lua was a "Tecgraf project" and Tecgraf served as "a good home" for it, according to Roberto. This continued through the late 1990s, even as Lua was increasingly looking outward. Around 2003, however, Tecgraf stopped paying for Lua development.

While this separation was to some extent expected, some of my interviewees attribute the ultimate break to the transition from Lua 3.2 to Lua

4. Released in November 2000, Lua 4 introduced substantial changes in the way Lua connected to code written in C. The change was originally motivated by the desire to allow a program written in C to run multiple Lua programs at the same time. Some of the users had requested this as early as 1998, and this ability turned out to be necessary for one of Tecgraf's own projects, called "CGILua," which aimed to make it possible to use Lua for developing web applications. (CGILua later became the basis for Rodrigo Miranda's Kepler.) The team originally introduced the feature by making the smallest possible modifications to Lua, trying to make sure that the new version ("Lua 3.3") would require almost no changes to the existing software. However, the limitations of this approach soon became clear, and the team proceeded to make more and more serious changes to how Lua connected to C code. Eventually the interface between Lua and C ("the Lua API") changed to a point where many existing programs would have to undergo serious modifications. The team then decided to use the opportunity to completely redesign the interface, thus changing it even further. When the new version was released in November 2000, it was deemed sufficiently different to be called "Lua 4.0" rather than "Lua 3.3."

While offering substantial improvements, the new API made obsolete all old C code interfacing with Lua. Roberto offered suggestions on how to fix the old code to make it work with Lua 4, but few Tecgraf projects undertook such migration.

Roberto: Then there was this big problem of compatibility. I think maybe this was the main breaking point. [. . .] The change from 3.2 to 4.0. That was a big change in the API so for people that only used Lua as a language, it was not that big, but for people that integrated Lua into other tools, the C API changed a lot and all applications in Tecgraf were in that kind of API stuff.

Yuri: Was that something you foresaw?

Roberto: The break or their reaction?

Yuri: Well, either.

Roberto: The break [in compatibility] for sure we foresaw, but their reaction, I think . . . We wrote some compatibility code and some things to help, but people mainly didn't use it, at all. [. . .] They never changed to Lua 4. So they started to drift apart from the Lua community. I mean, because everything was written in new manuals, and new discussions and new tricks and everything was evolving around Lua 4.0 and they were . . .

Even CGILua, the project that motivated the changes that eventually led to Lua 4.0, never released a version that worked with Lua 4.0. Unwilling to make the transition, Tecgraf's projects got "stuck" with Lua 3.2, a version that soon started to lose the interest for the Lua community. Lua's and Tecgraf's paths started to diverge.

In addition to the practical problem of backward compatibility, Lua 4 set the precedent for introducing features that brought only cost and little benefit to Tecgraf's projects. Some users of Lua did not take this well:[16]

Roberto: I think that they got . . . with some reason I think they got a little offended with the change to 4.0. I think that's why it was kind of a break[ing] point. I think this was the first change that we saw that it could hurt Tecgraf but we are going to do it anyway. We thought that it was not going to hurt *that* much, we tried and thought . . . Not something like "Oh, we are going to do it *because* it's going to hurt Tecgraf." We tried to minimize that, as I say, we did a lot of stuff to try to do compatibility layers and things like that. But we knew that it was going to have some problems, was going to be a big incompatibility.

Lua 4 was therefore not only introducing a technical break with existing Tecgraf software, but also demonstrating the new priorities of the Lua team. When I asked Roberto why this new API was introduced, he laughed: "Because it was *really* much better." "But better for who," I asked. "For any new user of Lua," Roberto explained. Making software better for new users in a way that hurts existing users can be a dangerous move in software development. However, the new Lua also worked quite well for its existing foreign community. Many of the list members were interested in Lua as a hobby and found the quick pace of change engaging intellectually. Others used it in games: software that is typically abandoned soon after it is released. (For example, two "versions" of *Grim Fandango* were released in 1998: 1.0 and 1.01. No other versions of the game were ever produced. Of course, some of the users may continue playing the game for years.) As far as foreign users of Lua were concerned, Lua 4 was a clear step forward and not only because of the improved API. Lua 4 demonstrated the authors' commitment to building a good language and their willingness to leave behind earlier mistakes. As earlier comments by Rich and Craig show, foreign users took note of this commitment.[17]

While some of Tecgraf's users of Lua complained about this transition, most of those I talked to in 2007 seemed to consider this sacrifice worthwhile. One early user of Lua said:

Silvio: It was in 4, I think, that the stack was introduced in the communication between Lua and C. And so whoever had much C code calling

Lua had to make lots of changes. And there were people who really complained: "Oh, darn, must change . . ." But I don't see it this way. I think we must keep moving ahead. Lua has to evolve. We are not going to stop and make Lua stagnate, or stay with an interface that we know is worse, just because there are people who use it and are feeling lazy to change their applications. It doesn't make sense. To stagnate for the sake of stagnation . . . Those who don't want to evolve can stick with version 3.2 and use it for the rest of their lives. It'll keep working, thank you very much.

In 2001, Lua was clearly showing global potential, and limiting it for the sake of Tecgraf's older projects did not necessarily make sense, even from the perspective of some of the Tecgraf engineers already invested in earlier versions of Lua. Lua's success abroad was starting to bring certain dividends to PUC and Tecgraf (in terms of prestige if not money), as well as individual people at Tecgraf who, like Silvio, were incorporating Lua into their academic research. Constraining Lua's growth was not necessarily in Tecgraf's best interest.

At the same time, Tecgraf itself was increasingly looking at other technologies. The rapid software innovation in 1990s meant that by 2001 Tecgraf was getting requests for new types of applications and could make use of new tools for implementing them. The most important of those was Java—a programming language released by Sun Microsystems in 1995 that had become the new standard by 2001. Even such committed supporters of Lua as Silvio saw those new technologies as better for some of their projects.

Silvio: It was about six years ago [in 2001] that we started increasing significantly the number of projects in Java. [. . .] In 2001 we had a request from the client, like, "Oh, we want a system with such, such and such characteristics." And I thought it would be more interesting to use Java than . . . Because there is this thing . . . There is this saying: "For someone who has a hammer, everything looks like a nail." We have to have a toolbox and to know when to use each tool, right? Lua is a great tool, but it's not the right tool for everything. Nobody would expect it to be. For the job that we had in front of us, the ideal solution was a mixture.

Despite Lua's oft-cited portability, Silvio's team found that using Java made it easier for them to run the application on their clients' computers without having to worry about which operating system the clients were using. Additionally, while Lua still worked best for certain Petrobras-specific functionality, Java offered simpler solutions for the more generic problems such as the construction of user interface. The team settled for a hybrid approach: the clients used a program written in Java on their desktops, which ran inside a web browser and connected over the network to a Lua program running on a remote computer.

While Silvio's explanation stresses the practicality of the mixed approach, it also alludes to the engineer's need to maintain a diverse "toolbox." Using Java also offered Tecgraf engineers an opportunity to gain experience with a new technology that was growing in popularity. For many, this dramatically broadened their options for employment if they were to ever leave academia and move into the software industry.

While Silvio's project started as a mixture of Lua and Java, it has gravitated toward Java over time. Those parts of the project that were done in Java ended up requiring more code than the parts written in Lua. "So today the project is mostly in Java," explained Silvio. Other projects started at the time were done in pure Java from the beginning. For those users of Lua less committed to the language than Silvio, the break introduced by Lua 4 served as a good opportunity to switch to Java.

* * *

This chapter looked at the history of Lua, seeking to show Lua's transition from its creation as a specific solution to a particular need of a Brazilian organization to its emergence as an international programming language that in retrospect may appear to have been "born to fly." As the authors themselves stress, this global success was neither planned nor even fully imagined. In many ways, it "just happened," with the authors' own understanding of the project and its possibilities shifting substantially over time. In some cases, Lua's trajectory was likely influenced by luck. It is important to understand, however, how the project found itself in a position to benefit from lucky circumstances.

Lua started as a "practical" project, aiming to solve specific problems for a specific client in Rio de Janeiro. This practical focus proved important, because it distinguished Lua from programming languages designed purely in pursuit of academic research. Had Lua been built specifically for the purpose of advancing computer science research, it likely would have suffered the fate of School, the research language on which Roberto worked in the 1990s and which has since been all but forgotten. Lua's success in the foreign software industry, however, did not grow out of a success in the local industry of Rio de Janeiro. Lua never found much use in local companies and remained largely invisible inside Petrobras. (And while Petrobras was a rather large client, I never heard foreign users describe Lua as "used by Petrobras.") This lack of strong ties to the local industry was probably a blessing: stronger contacts with the industry would have likely entangled the language in local relationships so closely as to make its later international success quite difficult.

Instead, Lua made its way abroad through linkages between Brazilian and American academic computer science, as well as those between American computer science and the American software industry. Unwilling to seriously pursue the option of commercializing Lua for the local market, the authors made early steps toward globalizing Lua and then made it a topic of academic papers, published both in Brazil and abroad. Their efforts benefited from their ability to explain their project in the right language—both in the sense of literal fluency in English, but also in the sense of fluency in the conceptual system of academic computer science. They had acquired this competence in part through physical travel to foreign centers of computer science research (such as Waterloo and Cambridge, UK), as well as through their access to a local island of computer science research in Brazil (PUC's Department of Informatics), which had been constructed in Brazil through the combined efforts of many people over several decades, as we saw in chapter 4.

Foreign publications brought foreign users, in part reflecting the strong linkages between the software industry and academic computer science in the United States. This in turn helped the authors discover the satisfaction of interacting with a large number of users of the software they wrote, who not only expressed gratitude but also had the sophistication to truly understand the virtues of the language—and to push its boundaries. Such satisfaction is of course similar to the one academics often seek when they publish their ideas with the hope that they would be valued by peers. The authors could thus acquire the "perceptions and judgments" necessary for the development of free software by starting with a rather similar set of academic "perceptions and judgments," and then building on them gradually while interacting with Lua users.

While Lua's ties to the Brazilian software industry were never strong and in many ways only weakened over time, this does not mean that the language has been entirely disconnected from a local system of economic relations. Lua's authors did not grow rich from their work on Lua, but their work did receive financial support in the form of salaries (paid by PUC-Rio, Tecgraf, and IMPA) and research grants from the Brazilian government.[18] It was the nature of this support, the fact that the authors were being paid not for their contributions to the organization's short-term profit but essentially for bringing it longer-term prestige, that made it possible for the authors to not have to worry about making money on Lua and to instead offer it as free software.

This chapter has focused on Lua's gradual disembedding from the local context and the increased success abroad. As we saw in the end, this success required breaking some of the local relationships. In the next chapter I turn more closely to this issue, looking at Lua's relation with the university, the city, and the country where it is based to this day. We will also look at some of the complexities involved in managing a global project from a university in Rio de Janeiro.

7 Fast and Patriotic

Nineteen years after its first version was developed, Lua is a fairly popular language, used in a number of well-known software products, both commercial and open source, with the number growing every day. The work on spillover effects in innovation may lead us to think that Lua's success would present an important opportunity for local economic development: local companies could take advantage of their proximity to PUC-Rio to gain better understanding of the language and its future directions, finding better use for Lua in their products and engaging in related innovation. This is not the case. At the time I was doing my fieldwork in 2007, Lua was largely unused in Brazil. Apart from Tecgraf, Nas Nuvens, and two other small companies incubated at PUC, Roberto Ierusalimschy knew of no Brazilian companies using Lua. If local companies were using Lua, they were not advertising this fact. Roberto remembered only a few occasions when local companies had entered into contact with the Lua team, none of which led to any extended collaboration. In my five months in Rio that year, I managed to find just one more company using Lua in Rio de Janeiro, bringing the total to five. By the end of 2008, three of those five companies were either moving away from Lua or had abandoned it altogether. The situation has been promising to change in recent years due to Lua's growing visibility abroad and its inclusion in the Brazilian standard for digital television, yet the language has yet to gain wide use in Brazil.

There are several reasons for this lack of local adoption. Some of my interviewees pointed, sometimes with much frustration, to the Lua team's seeming lack of interest in expanding Lua's use in Brazil. In fact, while Lua's authors mention in one of their articles being "bothered" by Lua's remaining relatively unknown in Brazil despite its growing use abroad (Ierusalimschy, Figueiredo, and Celes 2007, 2-9), I could see few signs of real efforts toward helping local adoption. At the time of my fieldwork in 2007, for example, Lua had no Portuguese documentation—an issue that did not seem to cause much concern for the team. A closely related reason is the seeming lack of fit between what Lua offers and the typical needs of the local industry. Lua provides clear value for two kinds of software projects: desktop software with high performance requirements (e.g., games) and small devices that cannot run the more popular programming languages. Both kinds of projects typically involve making *products*. Rio's software industry, however, focuses almost entirely on services to local organizations, which typically involves building web-based systems. Lua offers few obvious advantages in this domain. This lack of fit, however, can be seen as a symptom rather than a cause of the disconnect between Lua and the local industry. As we saw in the previous chapter, Lua's authors gradually adapted the language to the needs of foreign industry largely *because* of their lack of strong ties to the local industry.

Lua's disconnection from the local industry exemplifies a more general pattern of lack of ties between industry and academic research in Brazil, an issue often noted by my interviewees. There may be several reasons for such lack of ties. The main proximate reason is the government policy. Brazil's government funds academic research in accordance with the perceived academic success of each department and university. Such success is evaluated quantitatively and involves as an important component a metric of "intellectual production," measured by the number of publications. The publications are weighed by a rating that is assigned to each journal and conference by a government agency responsible for postgraduate education.[1] The rating goes from "A1" to "A2," then from "B1" to "B5," and finally to "C." Publications in journals and conferences rated "C" are normally given zero weight. The rating is based on each journal's position in citation databases such as Thompson Reuters's JCR, which primarily index English publications. Consequently, for computer science, Brazilian journals and conferences must usually include articles in English to get a rating above "C." Only those that publish articles *exclusively* in English get to "B2." None are rated "B1" or higher (CAPES 2009, 2011). Brazilian computer science researchers thus have good reasons to publish in "international" journals and conferences, which usually means those based in the United States. This in turn requires choosing problems that are deemed relevant by their American colleagues, whose interests often in turn reflect those of the American software industry.

A system of government funding that measured success by local use of research could shift this balance. The policy of giving incentives for publishing in foreign journals and conferences is not without merit, however, and its rationale aligns with the other, distal, reason for the disconnect.

Brazilian computer science researchers are located at the periphery of academic computer science. Brazilian software developers working in industry are similarly located at the periphery of their professional world. Each of the two groups must focus on building ties to the centers of their practice, both in order to keep their practice synchronized with the foreign models and in order to act as legitimate representatives of the practice locally. Successful publication in foreign journals provides Brazilian computer scientists the best way of demonstrating that their research is up to "world standards"—to funding agencies and also to each other. The use of this standard helps align the individual researchers' incentives with the *collective* goal of establishing the validity of computer science research in Brazil. Focusing on the needs of the local industry could lead the researchers away from the central problems of the "global" research practice. This would make it harder to be confident that Brazilian computer scientists are practitioners of global computer science who happen to be based in Brazil, rather than practitioners of some "Brazilian" computer science. This could in turn easily undermine their credibility in the eyes of the *local* industry.

Local firms similarly find it safer to stick with "standard" technology, choosing Sun's Java or Microsoft's .Net over PUC-Rio's Lua. Use of locally produced research is risky since the quality of such work cannot be easily assessed. Even in cases where the developers may believe in the technical superiority of the local product, choosing them may be unwise, as it may scare the clients. (See Fabio's discussion of Kepler in chapter 5.) This becomes especially so when the product in question claims to be innovative. "If it only exists in Brazil and it's not *jabuticaba*, then it can't be any good," says a popular Brazilian proverb. (*Jabuticaba* is a fruit tree that grows in parts of Brazil.) When a local technology does gain some local traction, it is common to assume that nepotism was the reason. Companies thus often fall back on the safer assumption that local technology can be ignored as irrelevant.

To understand the challenges of putting Lua to use in Brazil and of creating the kind of linkages that could help Lua bring about economic development, we would also have to look at the risks involved in linking Lua too closely to the place where it was born and the challenges of running a global project from the periphery. As we saw in the previous chapter, Lua's success was not planned by its authors and in many ways "just happened." This does not mean, however, that continued future success of Lua poses no challenges. In fact, it only makes the situation more complex for the authors, who do not fully understand the factors that led Lua to its current position and hence have a limited idea of what awaits it in the future.

While such problems may be faced by developers located at the centers of the software world, we will see that some of the challenges have much to do with the peripherality of Lua's authors.

A Little Bit of Actual Patriotism

Even as the use of Lua declined at Tecgraf in recent years, many of the original users of Lua, such as Antônio and Silvio, have continued to read the Lua mailing list and maintained an interest in the language. Some of them have contributed code libraries to Lua. In addition to saying that they follow Lua because they are currently using it in some projects, Lua users at PUC typically mention two factors contributing to their continued interest in the language: Lua's origin at PUC, discussed as a matter of sentimental or personal attachment, and the desire to see Lua attain the success that it deserves:

Silvio: I continue to use Lua in Tecgraf projects, so there is a practical reason [for following the list] as things happen in Lua, that's of interest to me. [. . .] So I want to be a part of anything that happens [on the list] because it may affect me in a project. The other reason is that I *love* this, I am an enthusiast of this language, I love Lua; I think it's awesome and I *like* to read the discussions; I actually do it because I *enjoy* it. So there are those two aspects.

When I asked Silvio what he meant by "loving" Lua, he explained:

Silvio: Yes, because I've accompanied Lua since the beginning, right? Roberto was my master's advisor. I mean: undergraduate, master's, and PhD. I have a friendship tie with him. Every now and then we go and have lunch together, he tells me what's happening, I don't know what. So I very much live in this world of this language and I do enjoy it, being a part, seeing what happens, talking to Roberto, exchanging ideas. So it's like this. It's something I like, really like. Because of the people involved. It's a *fantastic* piece of work.

Other members of the local Lua community at PUC often stressed the same reasons: Lua was an amazing piece of software demonstrating the genius of Roberto Ierusalimschy, a person who they considered a friend or a mentor; they felt a connection to Lua having seen it from the days when it was completely unknown. Many added that Lua also brought prestige to PUC—a university with which they were themselves affiliated. The fact that Lua had been developed in Brazil or in Rio de Janeiro was rarely mentioned.

In early March 2007, as I was reading through archives of *python-brasil*—a Brazilian mailing list dedicated to the Python programming language, I came across a thread entitled "Python–Lua." The thread started with a request for comments about Lua and a question about its advantages and disadvantages compared to Python. While such questions about "competing" programming languages sometimes invite hostile responses, most of the replies were quite positive. One of them read:

Python has a more elegant syntax, a larger community, a more diversified field of use, better interoperability.

Lua is faster, leaner, more patriotic, more adequate for "embedded systems" [in English] and has a VM for Palm [. . .].[2]

The next day, I mentioned the thread to Rodrigo Miranda, saying that "Python people" were discussing Lua on python-brasil. "They said it was fast and patriotic," I summarized the discussion. "It *is* fast," agreed Rodrigo. Noticing that he confirmed only half of the statement, I asked him explicitly: "Is it patriotic?" "I am not into this kind of stuff," Rodrigo responded.

While stressing their personal connection to Lua's authors and Lua's connection to the university, PUC users of Lua only infrequently brought up the fact that Lua was developed in Brazil as a reason for supporting it. When asked about this explicitly, some, like Rodrigo Miranda, explicitly denied any interest in "Brazilian" software, sometimes referring to such "nationalistic" sentiments as demonstrating narrow-mindness or even a lack of education.[3] Others admitted such feelings after some hesitation:

Yuri: But it's not a matter of Lua being a Brazilian language, I don't know . . .
Antônio: [Pause]. There is a little bit of this too. Because I know people who made it. I wanted to help promote it in some way. [Lists several reasons for promoting Lua.] But there is a little bit of this too, definitely. Of pride, of knowing where it came from, of promoting a domestic software product [*um software nacional*]. Definitely.

The Portuguese phrase Antônio uses to describe Lua in the end—"um software nacional"—is remarkably ambiguous. While it can be literally translated as "national software," such a translation would connote a lot more patriotic pride than the Portuguese word "nacional" typically implies. Products that are described with this adjective are often understood to be local substitutes for foreign products that are either not available or more expensive. (For example, a visitor to a Brazilian bar may be offered a choice between an expensive "whisky importado" and a cheaper "whisky nacional.") I thus use the more neutral term "domestic" to translate this word.

Considering the baggage carried by the term, it is not surprising that Antônio seemed uncomfortable saying that he supported Lua as a case of domestic/national software, mentioning this only when asked directly, and only after first making it clear that he had many other reasons for supporting the language. As in nearly all other interviews, the answer to my question about Lua as "Brazilian software" came only after a long pause.

For many educated Brazilians, feeling "patriotic" is quite appropriate when the Brazilian national football team is playing in the World Cup—especially in a game against its main rival Argentina. In many other contexts, however, "Brazil" is quite often a name for a package of problems that one must deal with rather than something to be excited about or to cheer for. In the context of technology, the focus on the national becomes particularly dangerous, because it suggests parochialism and an inability to grasp the values of the larger, global technical culture. A good engineer should not let his judgment be swayed by nationalistic feelings, say the developers. For those in their thirties and younger, who were consumers but not producers of technology during the market reserve years, Brazil's technological "backwardness" in 1970s and 1980s provides ample proof.

The perception of parochialism that casts its shadow over any local projects (and which can only be disarmed by personal acquaintance or international success) can be illustrated by a story told by "Ricardo," who was introduced to Lua in 1998 as a student at PUC:

Ricardo: I remember that we had to do a [class] project in C and she [Roberto's spouse, also a PUC professor] taught a new language, which just existed for a few years, invented at PUC and called "Lua." I looked at that and was like: "Eew! A language invented here at PUC? How stupid! I am not going to learn this. I'll never use it in my professional life! What will I do with it? And I remember there being two parts to the assignment that she sent us. One part was in C, another part in Lua. And a girl who was doing a part of the assignment with me . . . [I told her] "Here, do the part in Lua, because I am not going to learn this stuff, I don't want to know about Lua. I'll do the part in C, which is more interesting, since I'll use it."

Despite the fact that the class assignment introduced Lua in the exact context where it was strongest (as a part of a Lua/C combination), Ricardo avoided any contact with the local language, referring to the very idea of a programming language developed at PUC as "stupid" (and relegating the task to a woman).

Responding to such clear expressions of prejudice was about the only context in which Brazilian Lua users introduced—with hesitation—the topic of Lua's Brazilian origin.

Silvio: I think it's great that Roberto managed this . . . There is also this other thing . . . It's like . . . I already saw lots of prejudice against this language. It's impressive how this happens. Like this: I saw one of our clients, a person inside a company, say the following: "Oh no, I am here trying to decide whether to use Lua or this Microsoft application. But I think I will go with Microsoft, because if I use Lua and run into difficulties, my boss will think I am crazy. And if I use Microsoft and run into issues, that's not a problem, because this happens every day." It's *ridiculous* to think like this. I feel ashamed to see someone talk about it this way. I mean . . . Instead of being able to use . . . Instead of *promoting* work that was done completely in this country . . . [For] the guy not to promote it out of *pure prejudice* . . . I think it's totally ridiculous . . . So . . . Those kinds of things motivate me to follow Lua, to try to use it. Because I see that the list, most of the people on the list are not from Brazil, most are foreigners. I saw that Lua has to succeed abroad to gain acceptance at home. [Pause.] In other words, it's a project that I think is fantastic, which I really like. I see Roberto's struggle to make Lua work, I see the work he has to do. Lua reflects, deep down, his genius. And I think Roberto is very good. [Pause.] For those things, I really like the language and it's a pleasure for me to follow its growth. For this reason I don't leave the list and continue reading it. Even if I don't say anything, I stay on the list seeing what happens.

Silvio referred—only in passing—to the possibility of "promoting" the work done in Brazil. This course of action, however, was presented only in contrast to the prejudice that Lua had faced in Brazil over most of its history.

As I mentioned in the previous chapter, American users often indicate that they "take a global perspective" on things like Lua, concerning themselves relatively little with where such software comes from, as long as it is presented in good English in a way that shows technical competence on the part of the authors. As indicated by the earlier quotation from Ricardo and the story related by Silvio, Brazilian software developers often find that they do not have the luxury of such a "global perspective." Often working in contexts where their competence is questioned on a regular basis, they avoid diligently any associations that may bring accusations of parochialism. Unlike Rich, who became convinced of Lua's bright future by reading Roberto's book and noting the demonstrated global competence, potential users of Lua in Brazil may themselves lack the cultural skills needed to decide with confidence whether a programming language book written in English by a Brazilian author demonstrates the command of the global software culture or is a failed attempt to fake it. They thus find it safer to stick with tools whose global status is unquestionable.

While Silvio was disappointed by the "prejudice" with which Brazilians approached technology developed in Brazil, he realized that he had to accept some of the basic principles underlying this prejudice. To argue for Lua as a case of national software would be to invite yet stronger prejudice, feeding the suspicion that those who support Lua do so for reasons of narrow-minded nationalism. The best way to counteract the prejudice toward a homegrown language was to downplay Lua's Brazilian connections and look for global credentials that were valued locally. "I saw that Lua has to succeed abroad to gain acceptance at home," said Silvio.

Personal ties beat prejudice. A year after dismissing Lua as "stupid" in college, Ricardo joined Nas Nuvens, a startup working with Lua. At the time of our interview he was again employed as a Lua programmer for a PUC project, building software on top of Kepler. He talked about Kepler with enthusiasm, mentioning among other things "the idea of domestic technology":

Ricardo: I've always been following this Kepler thing and finding it interesting. Rodrigo would always pass by, like this, at PUC, and we would chat and he would tell me what was happening.
Yuri: But interesting in what sense?
Ricardo: [Long pause.] You see how things change? The idea of it being domestic technology [*tecnologia nacional*] . . . [pause], well-structured [pause]. A proposal for an actual framework for Internet development. To compete—perhaps Rodrigo would say that it's not to compete, but let's say that for now—with technologies that exist out there . . . I found it interesting.

The word "nacional" in Ricardo's phrase "tecnologia nacional" carries the same ambiguity as it does in Antônio's quotation above. I consequently again translate it using the word "domestic." Ricardo himself points to the difference between his current interest in supporting technology developed in Brazil and his earlier scorn for Lua as a student at PUC. This change seems much to do with his personal engagement with Rodrigo and other people working with Lua.

After mentioning the idea of "domestic technology," Ricardo quickly moved to the technical virtues of the project. I had to ask him to come back to this idea a few minutes later:

Yuri: What do you mean by the idea of domestic technology [*tecnologia nacional*]?
Ricardo: Huh?
Yuri: You said that a part of what made it interesting was this idea of domestic technology.

Ricardo: Oh, right. [Pause.] Why I find this interesting? [Pause.] It's hard to say. It's . . . [Pause.] I don't know really. Maybe a little bit of actual patriotism too. To believe that we [*a gente*] can develop really good, world-class technology. To be used by people from all corners [of the world], and which works.

Ricardo's informal "we" (*a gente*) appeared to refer to the Brazilians. When I asked him whom he means by "we," he confirmed my guess with hesitation:

Yuri: "We" means who?
Ricardo: I can say "we the Brazilians," or I can say, "we . . . PUC," "Nas Nuvens," "the open source community." I don't know, I don't know really. It's like this.
Yuri: But who did you have in mind?
Ricardo: I think that . . . [Pause.] This idea of domestic technology [*tecnologia nacional*], or let's say developed—if only initially—by people here, in Brazil. I think this excited me. Even if—"Okay, there are people from all over the world participating." Even better! Do you understand? There are people from all over the world offering recognition to something that started here. [Long pause.]

Ricardo struggled to bring together two seemingly contradictory ideas: Kepler and Lua as examples of "domestic" or even "national" technology (something Brazilians can be proud of), and the "global" nature of those projects. He arrived at a formulation that was similar to Silvio's prescription for Lua's success: technology produced in Brazil could be a cause of pride when validated by acceptance around the world.

For Everyone's Benefit

On my second day in Rio in 2007, I went to meet Rodrigo Miranda, to catch up on what had happened during the year I had spent in California. One piece of news that Rodrigo related to me concerned the recent interest that PUC had developed in Lua. In fact, Rodrigo said, there was now a project, to which I will refer here as "Iris," aimed to promote the use of Lua by setting up a number of publicly visible Lua projects (patterned after Rodrigo's own Kepler, described in the chapter 8) and looking for funding from foreign companies and local agencies. Rodrigo talked about Iris with excitement, even as he mentioned some of the internal politics around the project and the fact that the talk had been running somewhat ahead of the action.

When we returned to the topic a few days later, Rodrigo told me that a little over a year ago he presented the idea to "Chico," a friend of his, who

at the time was starting to exercise a certain amount of influence at PUC—
at least more so than Rodrigo himself. Chico liked the idea and went to
talk to his boss "Carlos." Carlos had heard of Lua by then, but did not yet
have a plan for how to use it to PUC's advantage. He liked Rodrigo's idea,
however, and took it to *his* boss. According to Rodrigo, Iris eventually made
it to the highest levels at PUC, and became "a big thing." As the project
grew in status, PUC managed to line up some money from "Softnet," a large
American IT company. Now that there was money involved, naturally even
more people were excited.

Rodrigo said he had discussed the plan with Roberto early in the process,
but Roberto perhaps had not taken it seriously, thinking it was another of
Rodrigo's "crazy" plans, unlikely to ever materialize. After the project grew,
however, Rodrigo told me, Roberto began to hear about it and complained
about not being in the loop. While this had caused some tensions around
Iris, Rodrigo was confident that this was a matter of miscommunication,
and was hopeful that the issue would get resolved, as Roberto would come
to see Iris for the opportunity that it was.

A few days later, Rodrigo introduced me to Chico, who greeted me with
much excitement. After a brief tour of his lab, Chico talked about his efforts
to "evangelize" Lua. *PUC had not been paying attention to Lua*, he said. *But
it's changing now.* At some point, he turned to Rodrigo and remarked: "I am
going to go to the Dean and say: Look, there is a guy from *the United States*
doing research on Lua! Why are people *here* not paying as much attention?"

Later that month, with some help from Chico, I met Carlos, who told
me of his reasons for supporting Lua. As he explained, he had known
Roberto for some time, as a friend of a friend, and had heard of Lua before,
but he had not thought seriously about its potential until recently, when
he started noticing its popularity abroad. "When I took this planning role,
and started looking for the potential of the university, what it had," says
Carlos, "I noticed that there was a fairly big movement actually using Lua."
Chico's return to PUC brought Carlos in contact with Rodrigo Miranda
and his ideas about strengthening Lua's local position. Carlos embraced
those ideas and took them to the university's administration, making it his
goal to reinforce the links between Lua and PUC, a tie he thought would
be beneficial for both the language and the university. Carlos referred to
his efforts as "evangelization." "You have to win people's hearts, people's
minds," he explained.

Stronger Lua would benefit not only PUC, but also the city and the
country, stressed Carlos. It could help curb the prejudice applied to local
technology. Unlike younger engineers and scientists, Carlos was not afraid

to come across as someone whose technical judgment was clouded by his concerns about the future of Brazil, speaking enthusiastically about the opportunity of local economic development that Lua created:

Carlos: The perception that if we can unlock the value of this language, it will serve for everyone's benefit, for Rio de Janeiro and for Brazil. Because a peripheral country always has this . . . Always: "Oh, no, this is just made by a Brazilian." You buy a product in Brazil, if it breaks you say: "Ah, it's Brazilian." You buy foreign equipment, a car, you say: "Ah, look how wonderful." If it breaks, the person almost doesn't even say anything. But if you bought a Brazilian product [and it breaks], you say: "That's because it's Brazilian." And this applies to software too. Despite our great position in banking software, in the financial sector, etc. . . . But Brazil is not very aggressive in this *offshore* field.

According to Carlos, Lua could change this perception, leading to new economic opportunities for the region.

This change, however, would require support from PUC's administration, policy makers, and funding agencies, explained Carlos. Local companies were unaware of the opportunity Lua offered them, quite likely due to the same prejudice. The few that understood this value found themselves unable to find Lua programmers. (Carlos seemed to suggest that the fact that local companies did not use Lua then in turn meant that there were few incentives for the programmers to learn it.) This could be changed with some support. "And what I also presented is that maybe we need to make an effort to look for resources," said Carlos. "To bring financing agencies here, tell them: let's train Lua programmers." Such training programs, combined with other forms of support to companies interested in offering Lua services to foreign clients, could jumpstart a new sector in Rio's IT market.

The initiative was to be presented a month later in a meeting with the secretary for economic development of the State of Rio de Janeiro:

Carlos: . . . where one of the proposals that we will bring for him for the development agenda would be the issue of information technology, but with a focus on Lua, as Brazil's differential for *offshore*. Java could be in China, in India . . . India has this ease with language, which here in Brazil—perhaps we do not have that. But without any question, Lua is a differential where today we have the conditions to quickly form a critical mass, if we can articulate this. [. . .] We'll invite financing organs, in the area of science and technology, the National Bank of Economic and Social Development, so that they come here, and bring the companies that are already in the process of religious conversion [*catequese*] . . . [Laughs.] For them to also talk about their interests. In other words, you would in a way be underlining and certifying that this is a Brazilian product, open source, and even [showing] this creativity here in Rio de Janeiro.

With some support, Rio software companies could start offering outsourcing services in Lua, benefiting from their proximity to Lua's base. As Lua would grow (in part thanks to this local business activity) and as long as its national origin is properly highlighted, its success could help change the perception of Brazil and Rio de Janeiro by displaying the technical creativity that can be found in the city.

While wanting to position Lua as "a Brazilian product," Carlos was also global in his thinking, feeling that Lua and PUC needed global partners to make Lua a true success. "Because if Lua has a great reach at the international level, if it has a series of qualities that the community recognizes, why not attract heavyweight partners, big international *players*, like Softnet, IBM, to talk about this topic?" he asked, using the English word "players." As a first step in this plan, Carlos had recently negotiated, on remarkably good terms, research and development funding from "Softnet," which was to go toward projects using a combination of Lua and Softnet's technology.

When I discussed the Iris project with Roberto Ierusalimschy, I learned that he was less than excited about the idea. While this had partly to do with his fear that Iris would end up competing with his own research group for "Lua" funding, he also feared the potential costs to Lua itself. "I'm not sure whether PUC has a clear notion of what it wants to do with Lua," he explained.

Even with the best intentions, the project could harm Lua. Legally, Lua's copyright belonged to PUC. If PUC were to enter into a contract with a big company, such a contract would need to be written and presented with much care to avoid creating an impression that PUC was giving their partner some kind of exclusive rights to Lua. Roberto seemed unsure that PUC's lawyers and administrators would know how to handle this. PUC had absolutely no experience with free / open source software, explained Roberto, and little experience with software licensing in general. Lacking understanding of how free software worked and how Lua fit in the world of free software (an issue that Roberto himself felt he barely comprehended, as we will soon see), PUC's administration risked creating the wrong kind of ties between Lua and the local context from which it was finally starting to separate itself.

Among other things, such contracts could highlight the image of Lua as software from "a South American country"—a product potentially

enmeshed in the bureaucratic complications that one could only successfully navigate by having strong local connections.

Roberto: Because one of the main problems Lua had . . . I think now finally it doesn't have it anymore, was that people were very unsure about
using a software from . . . as you say, from a South American country, they
don't know how we think and things like that. So they are afraid, "Can
we use Lua? Is it really free? What they are going to do?" Things like that.
You told me, you know that.[4] And then after all these years we kind of
conquered some kind of credibility. "Oh, okay, Lua seems to be something
stable. A lot of people use it, so I don't think . . . there is not going to be any
problem." So more and more people are starting to use Lua. And more important, more and more people are starting to *admit* that they use Lua. And
then suddenly it transpires that there is a hidden contract done in South
America between IBM or Softnet or Microsoft or whatever and PUC, which
is the legal owner of Lua, that . . . whatever it is in the contract [laughs], it
doesn't matter very much . . .

Since Lua had been released under an open source license (with the tacit
agreement of the PUC administration), the question of who owns Lua's
copyright could turn out to be quite irrelevant.[5] The *uncertainty* created by
badly written contracts (or simply bad publicity around contracts written
well) could be enough to scare off Lua's users. "Nobody wants to go to court
in Brazil to try to use Lua," said Roberto.

Concerned with unlocking Lua's value and using it for everyone's benefit, Carlos wanted to "reinforce the connection" between Lua and the local
context in which it was born. This association, however, was precisely what
had to be undone for Lua's global success. As Silvio said, to succeed in Brazil, Lua had to first succeed abroad. To succeed abroad, however, it had to
avoid any associations that could go with "software from South America."
Such successful detachment reduced Lua's ability to change the perception
of Rio and Brazil. But a slower, more patient approach would perhaps work
better in the long run. "What I would want to happen would be that PUC
got a better notion of the outside value of Lua before a notion of the inside
value," Roberto said. PUC could do little to enhance Lua's global position,
but could do much to hurt it.

Roberto's concerns with Carlos's efforts to get PUC to pay more attention
to Lua had much to do with his feeling that PUC's administration did not
have a clear idea of what it meant to run an open source software project.
Despite spending a substantial amount of time thinking about this topic,
Roberto himself was unsure whether he understood all the intricacies and

implications of open source. In the next section, I turn to Roberto's efforts to make sense of Lua's open source future.

Reading Smoke Signals

Despite Lua's permissive license and the active interaction between the authors and the Lua community, Lua never fully moved to the open source development mode: all changes to the Lua code itself have always been made by the three members of the Lua team, mostly by Roberto Ierusalimschy. Roberto explained this decision in part by the need to keep the language small:

Roberto: Yeah, because I think most other programming languages—open source—they are much more open than Lua. So they are . . . For instance, in Python or Perl, you have a lot of people that actually vote for changes and there are those kind of open decision, open source decision-making strategies and things like that. You can enter as a committer and you are promoted as a developer and then you have the right to go and there is all this hierarchy. And Lua is just the three of us . . . [Laughs.]

I asked Roberto if they had thought about moving to the same style of development as is used in other open source software projects. "We thought about *not* doing that!" he said laughing. As Roberto then explained (and many Lua users seem to agree), community-driven development works well for *adding* features but makes it harder to control the growth of the language, not to mention *removing* features. Since Lua is generally recognized for its minimalism and small size, it could perhaps benefit more from what Raymond (1999) calls the "cathedral" approach to software development (a carefully executed vision of a single architect or a small group) than from the bustling "bazaar" of open source.

Roberto then added another reason, however: his desire to maintain control over Lua's future.

Roberto: The other main point is that we really like—I mean, someone once said that in kind of a very aggressive way, not that aggressive, but . . . It's our language, I mean, we like doing it and it's . . .
Yuri: Someone said what? That it was your language?
Roberto: Yes, but it was kind of "Nobody has nothing to do with that, it is his language, he does whatever he wants and he doesn't care what people think about it" [says imitating aggressive voice] and I mean, I care a lot about what people think about it, but I really want to keep this privilege

of—this is the language I developed, I want to have the language the way I want it.

While Roberto's desire to develop Lua as he wants it could be understood as a case of an artists' concern for the purity of the work, it connected to a concern that came up frequently in the interviews: the risk of losing control over Lua.

There were two sources of this concern. One was the inherent uneasiness of Lua's position as an international programming language with a base in the wrong place—"software from South America" trying to make it globally. The other was the Lua team's limited knowledge or understanding of Lua's use abroad, and especially of the social dynamics of that use.

At the beginning of our first 2007 interview, Roberto suggested starting with a different question from the one that I first asked him:

Roberto: I think maybe start a little earlier? Because this is something that I was thinking today and something that I am always thinking. The main point is that we have a very, very rough idea of the growth of Lua and how Lua is being used and things like that. We are always kind of . . . I don't know if this is because we are in Brazil or if it would be the same if you were living in Silicon Valley, but my impression is that I always kind of try to read smoke signals to try to realize that there is a real growth, [or] there is no real growth.

While Roberto said this issue was always on his mind, it turned out that a specific event had led him to spend time thinking about the topic earlier that day.

As Roberto proceeded to tell me, over the course of recent months (from November 2006 to March 2007) Lua had enjoyed a dramatic change in its position in TIOBE TPCI—a popular ranking of programming languages based on Google queries. After spending a long time in the group of the "next fifty" (languages that TIOBE rated as among "the top one hundred" but not in "the top fifty," without assigning them individual rankings), Lua entered the top fifty in December 2006, and then started a slow ascent within this group. The March 2007 ranking came out the day before our interview. Lua had made a dramatic jump: to the twenty-fifth position. Since most of the software development is done in just a small number of programming languages, being in twenty-fifth place did not imply a huge market share and did not yet qualify Lua for TIOBE's "A level" designation (which it did reach in 2011). However, it most certainly put Lua on the map, leaving it steps away from the doors of the most exclusive group in software: "major" programming languages.

This new success was so substantial that Roberto laughed in disbelief as he talked about it:

Roberto: [Laughs.] I am not sure if the index is wrong, I mean, it's a very, very big jump. It jumped over twenty other languages in one month. Very strange. [. . .] It's very strange, I'm not sure if they are right. But the main point is I have no idea how we are climbing up, what happened in the world that put us that much [up].

The reasons for Lua's recent success were largely a mystery to Roberto, who had to read "smoke signals" from a distance to learn whether Lua was growing and what factors might have been contributing to its growth. He similarly had little idea if Lua's use would continue to grow, how quickly, and, if so, what this growth would bring.

Apart from the discussion on the list, publications and blogs were the main source of information. "You're always trying to understand what is going on," he said laughing. Roberto told me a story from 1998 when a well-known columnist writing about programming languages mentioned Lua, describing it as a small language with a small user base—just some "tens of thousands" of programmers using it (Laird and Soraiz 1998). While this user base was described as "small" by the columnist, the estimate far surpassed the Lua team's own guesses about the size of its community. Such occasional surprises made it important to pay attention to what was being said about Lua.

Roberto: I try to *acompanhar*, to follow blogs, and I have a link in my bookmarks. There is a search in Google blogs for "Lua" and "programming" or "games" and every day I go to see news about Lua in the blogs to try to have a feeling if there is something new happening or things like that. And sometimes I see: "Oh, this is interesting." But sometimes I see something that I get surprised: "Oh, there is . . ." It's difficult to . . .

Following Lua news from an office in Rio de Janeiro took some work.

While such uncertainty is inherent in the development of free software (since its use cannot be easily tracked through sales), especially when it is done on a small budget without the possibility of expensive market research, Roberto's situation was made more serious by his relative isolation from the place where his programming language was used most actively. Apart from the interactions on the list and the occasional Lua workshops (which by that point had happened three times, twice in the United States and once in the Netherlands), Roberto had very little contact with the programmers who use the language that he designed.

To understand the community around Lua, Roberto found that he had to get a better understanding of the "culture" of open source more broadly. He did so in part by reading a book about open source:

Roberto: Yeah, because there is this *culture*. For instance, I have this . . . I just . . . [Rodrigo] Miranda gave me a copy of this book called [. . .] *How to Run an Open Source Project.* It's assumed that an open source project is something that is open source decisions and . . . It even considers the possibility of . . . what they call . . . [pause] "the benevolent dictator." Is that the name? But it must be *benevolent*, that dictator. [. . .] And it's exactly because of that, he said in the book, because there is always this . . . the possibility of fork if the dictator is not benevolent, or if it's *really* a dictator. People will fork to another project and so . . . I mean, you cannot be a dictator . . .

As Roberto understood it, the author of the book (Fogel 2005; the actual title is slightly different) saw communal decision making (what Roberto calls "open source decisions") as a natural outcome of open source licensing. While leaders of open source projects are frequently referred to as "benevolent dictators" due to the influence they seem to exercise, their power is in reality quite limited. The reason for this is the inherent "forkability" of open source projects.

A free software license allows any recipients of the code to not only use or share it, but also to *modify* it as they see fit. Having modified it, they can make a case to other users that their version is actually better than the original. If the leaders of the project reject the modifications, the users can still make a choice to use the modified version. This choice potentially creates a "fork" in the development of the project, with some users sticking with the direction chosen by the project leader while others (perhaps very few, perhaps the majority) pursue the alternative paths. The possibility of forks limits the power of the leader. As Fogel (2005) explains: "Imagine a king whose subjects could copy his entire kingdom at any time and move to the copy to rule as they see fit? Would such a king govern very differently from one whose subjects were bound to stay under his rule no matter what he did?" (88).

As Raymond (1999) points out, in a section called "Promiscuous Theory, Puritan Practice," the right to create forks is essential to any free / open source license and such licenses can be said to "implicitly encourage" forking. One could say that the right to gain independence from the original author by "forking" a project is precisely what makes free software "free." In practice, however, forks are rare. As Raymond argues, this happens because of "an elaborate but largely unadmitted set of ownership customs" and a set of "taboos." Convincing the user community to abandon its current leader is consequently a hard task for the claimant, because it usually requires a combination of demonstrable technical prowess, communication skills, and a justification for what is often perceived as a clear violation of a social norm. Lacking those factors, the claimant will likely be ridiculed. (One derivative of Lua was subjected to such ridiculing by the mailing list members on a number of occasions in 2007 and 2008.) Consequently, project leaders often do exercise substantial power. In fact, the community's desire to avoid forking (which damages the community by splitting it) may actually enhance the power of the original authors as it leads the community to gather around the leaders and to constantly reaffirm their quasi-divine right to run the project.

The possibility of a successful adversarial fork that leaves the project leaders without the community, however, is ever-present. It looms especially ominously for leaders who may trust their own technical instincts but not their grasp of the *culture* of their users (in particular, all the subtleties of the unadmitted customs)—or find themselves unwilling to accept what they see as the demands of this culture. After reading the book, Roberto decided to take the risk and proceed with what he felt was best for Lua, disregarding what he understood to be the author's advice. He remained worried, however, about the consequences of that decision.

Roberto's experience with open source illustrates the challenges that open source presents for software developers at the periphery of the software world. While open source software development presents many opportunities for such participants, producing open source software successfully requires *higher* levels of competency in the software culture than other forms of engagement in software development. The developers working for Alta, who produce commercial software for their local clients, must only project a competent image to their clients and local peers. Their clients typically have limited grasp of the software culture and, in most cases, have few options for looking for expertise outside Brazil or even Rio de Janeiro. The authors of Lua cast their lot with a community of developers based largely outside Brazil and comprised of many people who are *more* fluent in the software culture than are Lua's authors themselves. This made a firm grasp of the global software culture crucial for their continued success.

Ginga and Beyond

When I returned to Brazil at the end of 2008, the conflict around Iris seemed to have been largely worked out. Carlos appeared to have understood some of the challenges faced by Roberto. He started consulting Roberto more closely on PUC's approach toward Lua, and the ambitions of the Iris project were largely scaled down. On the same visit, however, I learned of a new development that promised to substantially increase Lua's use in Brazil: the language had been included in the Brazilian digital television standard, as a part of "Ginga," a "middleware" component based on research done by another PUC team. Walking through the hallways of the Department of Informatics, I saw a poster advertising a course on digital TV programming that included an introduction to Lua. The existence of this "Lua course" at PUC was important news, relayed to me diligently by a number of my former interviewees. The course used the Portuguese translation of the Lua manual, completed by one of Roberto's PhD students in August 2007. After the interview, I passed by LabLua to chat with some of Roberto Ierusalim-schy's students. I saw *Programming in Lua* open on one of the last chapters. The same student who had translated the Lua manual was finishing his translation of the book, paid from Kepler's grants. Later that day I gave a well-attended talk about my own Lua project, the wiki engine that I started developing in 2007. The talk was organized by Chico, whose research lab dedicated to running Lua on small electronic devices had by that point expanded to several rooms. Local efforts to promote software based on Lua seemed to be starting to add up.

The rest of 2009 saw additional steps toward development of a local community. In late May the Lua team announced its decision to have that year's Lua Workshop in Rio de Janeiro. An email sent to *lua-l* announcing the event stated—in Portuguese, followed by an English translation—that in addition to the traditional goal of bringing together the Lua community, the workshop had another important objective: "disseminate Lua in the community and the industry in Brazil."[6] In August Rodrigo set up, with the blessing of Lua's team, a Portuguese mailing list *lua-br*, which had nearly a hundred subscribers after just a few days. (Today *lua-br* has close to 700 subscribers, while Lua's main list *lua-l* has slightly over two thousand; *lua-l* receives nearly ten times the number of messages of lua-br, however.) Such developments were perhaps somewhat stimulated by my own work—earlier that year I had shared with Roberto, Rodrigo, and Luiz Henrique drafts of my thesis, which included a version of this chapter.

In the two years that followed, Lua's presence in Rio continued to grow but at a seemingly slower pace. While Lua is now available in many television sets sold in Brazil as a part of Ginga, the standard also mandates support for Java. Brazil's main TV network Rede Globo appears to have decided to go with Java. (Adoption of Java-free version of Ginga, however, is being considered by Argentina.) The Portuguese translation of *Programming in Lua* has yet to come out. Lua's place in Rio remains somewhat uncertain.

* * *

This chapter explored Lua's complicated relationship with the city and the country in which it was born. As I argued in the previous chapter, Lua's global success had much to do with its successful disembedding from the local context. This disembedding presents both challenges and opportunities for the local use of Lua. Local developers must master a foreign language to use Lua. They also often find that Lua's strengths do not apply to the problems they face. There is another important barrier to local adoption, however, which has little to do with Lua's disconnection from Brazil: the developers desire to stick with languages more established on the global scene and to avoid suspicion of narrow-minded parochialism. In that case, Lua's distance from its local context in fact becomes an *advantage* for local adoption, because it helps Lua gain global credentials that are crucial for its local success. As Silvio said, he realized that "Lua has to succeed abroad to gain acceptance at home." Lua can bring benefits to Brazil (and already has, by contributing to Brazil's image as a place of innovative software work), but this would require continued success abroad and careful management of its linkages to the local place.

Silvio's realization reflects a pragmatic recognition of Rio's peripheral position in the world of software. In the next chapter, I explore the relationship between local and global innovation in the context of Kepler, Rodrigo Miranda's project that aimed to build a web development framework based on Lua—a different attempt to bring Lua home to Rio de Janeiro, one that in many ways attempted to reject (or transcend) the limitation of Rio's peripheral position.

8 Dreams of a Culture Farmer

"World domination," Rodrigo said with a smile, his tone suggesting that the answer should have been obvious. I had just asked him to clarify what he had in mind when he jokingly raised the question of whether knowledge of Chinese or Japanese would be more important for "his purposes." It was a Wednesday afternoon in May 2007 and we were at an *a quilo* restaurant in Copacabana, a typical Rio lunch place selling food by weight—a quick and relatively inexpensive option, popular among office workers. A block away was the office high-rise where Rodrigo's company occupied a few small rooms and where both of us had been spending most of our time for the last few months, sharing a tiny office, the backs of our chairs almost bumping into each other. It was a somewhat odd place to be plotting world domination. We laughed. A few minutes later we got up, paid (making sure to ask for a corporate discount), and headed back to the office. We did have a world to conquer, after all.

Rodrigo, whom the reader has met on a few occasions in the earlier chapters, was a Carioca in his late thirties who had dedicated the last decade of his life to building a web development platform based on Lua, the programming language discussed in chapters 6 and 7. Since 2005, Rodrigo had pursued this goal in the form of an open source software project called "Kepler." I first met Rodrigo during my time in Rio in 2005; he was, in fact, my first technical interviewee that year. While finding the project intriguing, I dismissed it as an outlier, choosing to dedicate my time to interviews with developers working in more typical companies, not unlike Alta. I stayed in touch with Rodrigo, however, and as I learned more about Kepler over the course of the following year, I gradually came to the conclusion that observing a project aiming to develop a global product based on local innovation would provide a useful contrast to a study of a "typical" company, focused on bringing foreign technology to Brazil. When I returned to Brazil in 2007, I decided to accept Rodrigo's invitation to study Kepler and his offer of a desk in his office, using it as a base for my investigation of Lua as well as for looking at Rodrigo's own project—the subject of this chapter.

Kepler and the company that sponsored it stand in contrast with both Alta and Lua, adding important complexity to our exploration of the world of software. In contrast to Alta, Kepler represents an attempt to create a global platform—a collection of software developed for the world rather than for the needs of specific local clients, software meant as a foundation for other software development rather than a product in itself. While Rodrigo's talk about "world domination" was a joke (as were his colleagues' references to him as "the Brain" of "Pinky and the Brain"), the project did have global aims, positioning itself in competition with globally popular web development technologies. Its audience lay primarily outside Brazil.

Unlike Lua, however, whose success can be attributed to a combination of preexisting foreign ties and successful disengagement from the local context, Kepler had to create crucial local alliances, on which it came to rely. One of those alliances involved betting on Lua itself—a programming language developed just a few kilometers away from Rodrigo's office, by his former advisor. Rodrigo had also decided to work together with a local company oriented toward Brazilian clients, using local developers, and drawing on funding provided by the Brazilian government.[1]

Paradoxically, local use of local innovation represents a starkly global move. Using local innovation is something software companies in Silicon Valley do all the time. It is not commonly done in Brazil, however, where foreign technology is usually seen as a much safer bet. Rodrigo's project was thus simultaneously global and local in important ways—"glocal" to use Wellman and Hampton's (1999) term. It drew on Rodrigo's imagining not only a global technology platform but also a transformed local place. This chapter shows some of the challenges involved in pursuit of such glocal dreams at the periphery of a global world of practice.

Attempting to bring together the resources of local universities, the local industry, the national government, and the remote world of practice in pursuit of local innovation that aims to be global in its significance, Kepler represents the kind of innovation that may be crucial for development. Understanding its challenges may thus provide valuable insight into the dynamics of technological development at the periphery.

Making Waves

Rodrigo Miranda grew up in Rio, the son of an engineer and a journalist. As a child, he wanted to become an architect, but an encounter with a computer at age twelve led him to a new passion. By the time he was choosing a college program, Rodrigo knew he wanted to study computer science. He was planning to apply to the public Federal University of Rio de Janeiro (UFRJ), but PUC-Rio opened a new undergraduate program in computational engineering at the last moment.[2] Since his mother was teaching at PUC at the time, Rodrigo managed to get a scholarship to study there.

Rodrigo sometimes describes Kepler as the third of the three "crazy ideas" that he had pursued over the years. The first, an idea for a video game, had to be put on hold back in college. Rodrigo did not believe he could develop it by himself and could not convince his friends to join forces. The second vision, a hypertext database system, was abandoned in 1994 when Rodrigo saw the Mosaic web browser. The web, however, gave Rodrigo his third idea, one he had pursued for about a decade when I arrived at his office in 2007: a web development platform based on Lua.

Lua is today often used in computer games, a domain for which it is seen as being particularly fit because of its performance and ease of integration with C. Few software companies in Rio de Janeiro, however, build such products. Instead, most provide software services, which typically involves development of web applications. Lua is rarely seen today as a good foundation for web development. Yet web development in Lua has a long history. The first attempts were made in 1995, by a group of PUC students. It was "a game among friends," explained one of the participants, arising simply from a desire to understand the emerging web technologies. Little by little the project grew features, however. It was then picked up by another student, who developed it into a master's thesis and published several papers about the project (e.g., Hester, Borges, and Ierusalimschy 1997). CGILua, as the extended system came to be called, had a number of advantages over the better known alternatives, and in particular over writing web applications directly in C, a common practice at the time. While the first iterations of many currently popular methods for building web applications were being developed around 1996, none were really well known. For a short time, CGILua was perhaps about the best approach one could use for building web applications.[3]

At the same time as CGILua was being developed at PUC, Rodrigo quit his PhD program there and went to work as a manager in a web development company, which at the time built web applications in C and had a strategy not very different from Alta's. Being acquainted with the students who were working on CGILua, Rodrigo understood the time savings that CGILua offered over building applications in C and proposed using it for some of the client's projects. Rodrigo told me his manager refused to even consider the idea, finding Lua to be a toy language. Rodrigo proceeded to introduce Lua in the company surreptitiously, while also starting to think about the possibility of building a complete web development platform based on Lua.

While Rodrigo was looking for opportunities to pursue his technological dream, his brother "João" was pursuing his own. Unlike Rodrigo, João saw himself as an entrepreneur. His dreams, therefore, focused on starting and running a successful business. In 1997, having earned a small amount of money in a venture that provided software localization services to foreign software companies, João wanted to move to a new level: launch a *product* company based on proper development methodologies. In the earlier chapters I have referred to the company as "Nas Nuvens," a self-deprecating pseudonym suggested by Rodrigo and reflecting software developers' penchant for puns (much like the name "Lua"). "Nas Nuvens" means "in the clouds" in Portuguese, a name that suggests at the same time a degree of disconnection from reality (much like its English equivalent) and an association with Lua through an allusion to the phrase "no mundo da Lua" (literally "in the moon world"), a more idiomatic translation for the English "head in the clouds." (The name likely captures Rodrigo's doubts about whether the venture was ever realistic, but should not be interpreted as signifying my own position as to whether the efforts were worthwhile.)

João wanted to start a US-style company, hoping to distinguish himself from the local competition by a strict adherence to foreign software development methodologies and, as far as possible, foreign business methods. In a particularly stark attempt to follow foreign methods, João decided to build his company around local research—a practice standard in California but one rarely pursued in Rio. Having heard Rodrigo's praises of Lua, João decided that Lua could prove to be a key strategic advantage for a software firm based in Rio: close access to the Lua team would help the company stay abreast of any changes and perhaps influence Lua's development to its own advantage, while PUC would provide a steady stream of interns and employees skilled in the use of Lua. João also decided to start with a focus on the needs of the domestic market, perhaps expanding internationally over time.

As João saw it—correctly, I think—the local market needed an easy way to build web sites with dynamic content. Rodrigo's experience with CGI-Lua had proven that this could be done with Lua. Using his ties to PUC, João acquired a web publishing system based on CGILua that was built by another PUC student, using it as a starting point for Nas Nuvens's future product. He also managed to find startup capital from among relatives, friends, and a PUC professor. Soon after, Rodrigo joined the company as a chief software architect, seeing the venture as a chance to take CGILua to the next level and build a web development platform based on Lua in pursuit of his own technological vision. The two brothers were pursuing two different global dreams—one looking to create a global platform with relatively little interest in the financial aspects of the venture, the other looking to build a high-tech company using a foreign model, less interested in the details of technology to be built. Their parallel pursuit of their professional dreams, attempting to reproduce the practice centered in the same place, allowed for an alliance, much like many of the alliances described in chapter 4, and more generally illustrating the notion of parallel re-creation of practice that I discussed in chapter 1.

The company's product was innovative for the time and the company successfully obtained small innovation grants from FINEP, an agency that funds industry research.[4] João's superb skills as a salesman then allowed him to quickly attract substantial capital (about a million reais), hire developers, and actually build a product, something few entrepreneurs could do in Brazil even at the height of the dot-com boom. As the product was nearing completion, however, João faced a new challenge: finding additional money to actually *launch* the product, another two million reais in João's estimate. In 2000, as the clouds were starting to gather over the Internet industry, attracting additional money turned out to be a challenge even for João. Banks he turned to asked for astonishing interest rates, João told me. A large foreign company offered to invest, but did not want to do so alone. João eventually found what appeared to be a solution: the Brazilian Development Bank (BNDES) offered to fund the project. João knew that the bank would take months to make a final decision, but decided to take the risk, seeing few alternatives. When the offer got finalized, however, it came with a stipulation that the investment would be treated as a loan, to be personally guaranteed by João, which the bank would be able to convert into equity later if it so chose. Such terms left João with a risk of accumulating millions of reais in personal debt while giving the bank most of the profit in the case of success. In João's view, this perverted the meaning of "risk capital." He decided to pass on the offer and accept that a key ingredient of the California startup recipe—sensible funding—was missing in Rio de Janeiro.

Nas Nuvens was soon out of money and had to lay off many of the developers, including nearly all of the most skilled ones, who were also the most expensive. In California, this would be time to close the business, do some postmortem analysis, and then try again with a better idea. Brazilian bankruptcy laws, however, João explained, did not make this an easy option. João decided to keep the company running at a reduced scale. He ended the lease on the two floors that Nas Nuvens used to occupy, limiting the company to the few rooms where it has been since. The company shifted its focus to providing services to the customers it already had—an approach that many consider the only realistic route for a software company in Rio de Janeiro anyway.

Even this option, however, was turning out to be problematic. Unlike its competitors, Nas Nuvens had developed expertise in building web solutions using Lua and thus had to rely on CGILua, a rapidly aging platform that could not provide much of what the clients were starting to expect. The author of CGILua had by then departed for the United States and the PUC Lua community had largely abandoned the project, since it was looking to solidify Lua's success in other domains. (João's expectations of steering Lua's development had proved unrealistic—a tiny local venture was too insignificant, considering that Lua was being adopted by major multinational companies such as Adobe and Microsoft, whose needs were quite different from those of Nas Nuvens.) Shifting toward services and a closer relationship with local clients was also making the job less and less attractive to Rodrigo, who had himself abstained from financial involvement with João's venture and could leave at any time to pursue opportunities elsewhere.

The solution came in the form of government support. By 2001, FINEP showed less interest in the company's product, no longer considering it particularly innovative, as numerous foreign alternatives were becoming available. In 2002, however, FINEP started providing additional funding earmarked for open source projects, as part of the Brazilian government's broader push to promote open source software that was mentioned in chapter 4. It was thus decided that the company would apply for FINEP funding for an improved web platform based on Lua to be released as open source. FINEP's money and the allure of "open source" would make it possible to hire skilled developers from among PUC students and alumni and would satisfy Rodrigo's desire to pursue his vision. It would also, perhaps, attract contributions from abroad. Nas Nuvens would get a better Lua platform, which it could use to provide more competitive services. The project would be called "Kepler." As the project's web site explained, the name alluded to Johannes Kepler's discovery that tides on earth were caused by the moon, and was to suggest that Lua ("moon") was about to cause some tides.

The funding provided by FINEP was expected to be rather modest, and the company could not use it to hire as many qualified programmers as it would need. Rodrigo's connections to PUC networks, however, helped him find students and recent graduates already interested in Lua and willing to dedicate part-time efforts for less than the market wage. (As many of them noted, the fact that this was now an open source project was making it more attractive.) One of the new project members was "Márcio"—a recent PUC graduate working as a programmer, on a PUC web application based on CGILua. A dedicated fan of CGILua, Márcio had worked privately to improve the platform, without advertising his efforts. He had little interest in Rodrigo's plans for "world domination," but Rodrigo's offer of paying him to work on Kepler provided a good opportunity to dedicate more time to improving CGILua by not having to take on other consulting projects. Others joined the project partly out of interest in Lua and partly as a favor to Rodrigo, as we saw in chapter 5.

Rodrigo also returned to the Lua list and resumed his efforts to transform the Lua community to fit his needs. As Lua was typically used as a programming language embedded inside a larger C application—taking advantage of Lua's small size, efficiency, and ease of integration—Lua's community had traditionally emphasized minimalism and use of highly customized solutions. The members of the list had thus often preferred to share ideas rather than actual code, stressing that one project's code would rarely provide a perfect fit for another project's needs. Development of web applications, on the other hand, required a large collection of software modules that could be readily used. (In chapter 6 Craig specifically mentioned the fact that he was *not* trying to build something like a web server as a reason why Lua seemed appropriate for him.) Realizing that he could no longer hope to fund the development of all the modules he needed, Rodrigo attempted once again to convince the Lua community of the importance of sharing code and assembling a larger collection of modules. The culture of the Lua community was not quite the "right" culture for open source development, Rodrigo felt, but that was the community he would have to rely on. He referred to his efforts as "culture farming," implying a long-term investment in "growing" the culture that would then benefit Kepler. Since Kepler's own software was now going to be released under an open source license, Rodrigo used this as an opportunity to "seed" the sharing culture, actively announcing the release of each Kepler module on the mailing list and setting up a web site for sharing modules that he called "LuaForge" (a name chosen by analogy to that of a popular web site called "SourceForge").

When I first met Rodrigo in 2005, Kepler was relatively well funded and Rodrigo had a number of skilled developers working on the project. A later gap in funding decimated the team, as most Kepler developers felt that working for free was a luxury they could not afford—even though some of them perhaps wished they could. Márcio, one of the three developers who remained, took a somewhat different approach: continuing his involvement in the project but refusing to take money for it. This gave him the freedom to pursue other sources of income without constraining himself with a commitment to Kepler, which was too unreliable as a source of income.

While the development had slowed down substantially, the work done in 2005 was starting to pay off and the project was slowly attracting some users. In the spring of 2007, the project's mailing list (run in English, of course) had around one hundred people, who were increasingly involved in the project, at least to the point of asking occasional questions. Additionally, the Lua community was gradually warming up to the idea of sharing modules. LuaForge included around two hundred projects, with several new projects getting registered in a typical week. This was giving Rodrigo hope that the project perhaps could succeed. "My friends used to call me crazy, now they just say I am insane," he explained to me.

Opening Kepler

On one of my first days in Rio in 2007, Rodrigo told me he had a habit of going to PUC roughly once a week, to talk to the two Kepler contributors who worked there—Márcio, the PUC graduate who was working in PUC's IT department, maintaining an administrative web application based on CGILua, and "Tiago," a PhD student of Roberto Ierusalimschy. (A third major contributor, "Alan," used to be there as well but had just recently moved to Porto Alegre, about a thousand kilometers southwest of Rio de Janeiro.) Occasionally, he would also meet with Roberto, Chico, and others members of the local Lua community. A week later I was in a taxi with Rodrigo, going to PUC where I expected to sit through his typical meetings, observing the routine.

As soon as I got into the taxi, however, Rodrigo handed me a printout of a web article, in English, entitled "Financing Volunteer Free Software Projects" (see Hill 2005). I scanned it briefly. The article argued that paying open source developers was a dangerous practice, since paid labor would tend to "crowd out" volunteer contributions. In other words, paying some people would make others less willing to work for free, and even those that were paid might work less, since they would now see their work as an economic transaction. This article, explained Rodrigo, had helped him identify the problem that had plagued his project for a long time. He was releasing his source code under an open source license, but was not running

Kepler as a "real" open source project. He was hiring developers with the money provided by FINEP, and they worked while they were paid. When the money dried up (as was the case for most of 2006), the work stopped. Plus, he simply could not hire all the developers he needed: FINEP funding was limited and could only be used to pay people in Brazil. He had hoped to get people from other countries to participate without pay, but this had not happened. He had to get them involved. Perhaps this would require that he stop paying his current developers, even though the project depended on them.

We arrived at PUC and went to meet Tiago, a computer science PhD student Rodrigo paid to work on Kepler a few hours a week. We found Tiago in a small room that he shared with three other students. The three of us went into a conference room across the hallway and sat down under a whiteboard covered with set-theoretic formulas, half in English and half in Portuguese. After some brief introductions, Rodrigo announced that he had something important to discuss. He wanted to start running Kepler as a "real" open source project, he told Tiago. Kepler was a project with open sourced code, he said, but a closed development method. He wanted to change this. Perhaps the fact that Alan had just moved to Porto Alegre would make this easier: Alan's departure already made it impossible to resolve all the issues in face-to-face meetings at PUC, and they had to discuss most of the decisions by email and instant messenger. Now they just had to take it a step further. He wanted to start discussing more things on a mailing list, openly, explained Rodrigo. Of course they would use English for those discussions. Kepler already had a mailing list operating in English, used primarily for announcements. This list could become the center of the Kepler project, the new forum for the discussions that up until now occurred inside PUC walls. Tiago listened to Rodrigo speak, occasionally asking clarifying questions.

"This finishes the Weird Ideas of the Week part," said Rodrigo finally. "Now the practical part." Rodrigo and Tiago spent the next half hour talking about specific problems with the launch of Kepler 1.1. Even so, Rodrigo's suggestion of changing the model repeatedly seeped back into the discussion. They needed people to test, Rodrigo said. Perhaps people at Nas Nuvens could help, but this was also the sort of thing that mailing list members could assist with. Perhaps there should be a list of tasks on the web site where people could go and see what needs to be done. Rodrigo gave an example: a new person had appeared on the mailing list, asking what he could help with, but Rodrigo did not know what to tell him. While yearning for outside participation, the project was not organized so as to take advantage of it. There should be a page with tasks, said Rodrigo, some of them in red. In this case, they could tell new people: go to that page, pick a task.

We walked Tiago back to his room, chatted briefly with other PhD students there (as I later found out, Rodrigo was courting one of them to work on Kepler), then headed to meet Márcio, the other contributor. We were going to meet Márcio outside, near a kiosk that served coffee. While waiting for Márcio, I asked Rodrigo about the move to English. It was simply pragmatic, he explained. He had to draw on developers outside Brazil, so he needed a mailing list in English. He did not have the time to maintain two lists. Brazilians interested in Kepler would know how to read English and usually would know how to write it as well. Plus, Kepler had almost no documentation in Portuguese, so there was no meaningful way of engaging with developers who did not read English. Requiring Kepler contributors to discuss all the project plans in English on the mailing list would perhaps inconvenience them somewhat. Rodrigo was confident, however, that they were capable of doing this and would agree to do it if asked. When Márcio arrived, Rodrigo presented him with a much-condensed version of what he had said to Tiago. Márcio nodded, apparently finding no problems with this. Seemingly contradicting the new plan to move decision making from face-to-face discussions to the mailing list, Rodrigo and Márcio then proceeded to quickly discuss the state of specific subprojects.

Sometime later I had a chance to talk to Márcio about the proposed changes. He seemed divided. He did not share Rodrigo's desire to conquer the world, he explained. He did not care if Kepler was popular, he said. He wanted it to be *good*. That was the difference, he repeated: "Rodrigo wants Kepler to be popular, I want it to be good." When Márcio started working on Kepler, he did not expect it to be successful—it was a fun project that also paid some money. Now it had succeeded, however, and Márcio hoped it would succeed more. It was good for the world to know, he said, that here in Rio de Janeiro they were doing something interesting. For this reason, he supported Rodrigo's plan. He agreed that further success would require opening the project to outside participation and moving all decision making to the English mailing list. Doing so meant more work, but the effort was worth it for Kepler's success. But sometimes, he then added, he found himself just deciding not to write.

A week after my first visit to PUC with Rodrigo, we were again at PUC, sitting at a picnic table not far from the Department of Informatics, together with two other people. One was a potential future contributor Rodrigo was courting at the time. Another one was Renato, a close friend of Rodrigo who was always willing to lend Rodrigo an ear when Rodrigo needed a sympathetic listener. Rodrigo talked again about the idea of moving to an open source model and getting people on the list to contribute more. He did not really know how to do that, he said, since he had never run this kind of project before and his knowledge was only theoretical. "I have read books, articles on the web, talked to Yuri," he said, naming three foreign sources of information. But it had to be done. Kepler could not just rely on the efforts of local developers funded by FINEP. FINEP's funding was unreliable and paid only for new development, not for the work involved in keeping the software up to date, such as fixing bugs reported by the users. He had to find people who would be willing to contribute even without pay. And he had to look for such people outside. This "outside" (*lá fora*) was potentially ambiguous, as it could mean either outside the project, or abroad. In this case, however, the two largely coincided. Looking for help outside the project meant looking for it abroad, through Lua's mailing list. The theory of open source suggested that this could work. In practice, he did not know how it would work out. As Rodrigo talked, the rest of us just sat listening and asking clarifying questions. Our job was to help Rodrigo feel that his plan was not altogether crazy.

When we got up, it was clear that the decision had been made. The next day, an email to the Kepler list informed subscribers that the list was now going to function as a "real" open source project. The lengthy message, entitled "Opening Kepler," started as follows:

Hi,

This mail got a lot bigger than I imagined at first... :o)

As you have noticed, the conversion of Kepler to Lua 5.1 is taking a lot longer than expected.

Not only we have found that this involves more work than we assumed, but also that the development model being used until here is not working as well as we would like.

The Kepler team is currently using a model that involves too much communication outside the public channels (this list and the site for example). We are trying to educate ourselves in order to change that, but this is not exactly easy for a team used to rely on interpersonal communications.

After laying out a detailed plan for specific Kepler components, Rodrigo concluded by saying:

As I hope you can notice by this mail, we are trying hard to move to a more open development model. That includes using this list in a different way and opening the site wiki for others to contribute.

Before going on, we would like to know what you think about the general idea and what other ideas could be added to this "new vision."

Thanks in advance for any suggestions and thanks for reading that much...

Rodrigo

The three subscribers who responded to Rodrigo's message expressed support for the plan. For some of them, Rodrigo's decision to "open" Kepler may have seemed obvious and natural. For Rodrigo, however, it represented an important transition and an entirely different way of organizing the labor that was to go into his vision.

Working the List

When we came back from lunch on that May Wednesday, we turned our attention from Rodrigo's long-term plans for "world domination" to the task at hand. Rodrigo and I were working on a chapter for an upcoming book about Lua programming. The chapter presented an opportunity to promote Kepler, putting Lua web development side by side with such accepted uses of Lua as embedding it in games or running it on microcontrollers. Our chapter focused on implementing in Kepler a "model–view–controller" (or MVC) application. MVC was a popular approach to organizing software applications and had been growing in popularity in web development for the last few years. Allowing MVC development in Kepler was one of Rodrigo's projects for this year. This goal, which seemed natural to me, was not shared by all of Kepler's contributors, Rodrigo complained. Márcio, in particular, felt no need for it, preferring the methods that Rodrigo considered outdated. Rodrigo talked about having to explain to Márcio that he had to implement MVC "to please his friends." Unsure whether the phrase referred to the foreign users of Kepler, on the mailing list, or to people in Rio, I decided to ask. Being in perhaps too jovial a mood, I phrased the question in a somewhat unfortunate way: "Are you talking about your real friends or your imaginary friends?" Rodrigo was taken aback by the question.

I do have a real friend, said Rodrigo. He mentioned Renato—a close friend who was with us at PUC when Rodrigo was finalizing his decision to open up the project. Renato was always willing to listen and be supportive. Perhaps too much so. The problem was that Renato would say "This is very interesting" to almost anything. Renato was too busy with his own job to follow Rodrigo's interest in depth. He was willing to express support, but had no ability to accompany Rodrigo in the journey. Rodrigo then turned back to Kepler contributors—not exactly friends, but perhaps colleagues, he explained. For some of them, such as Tiago, Kepler only made sense with MVC. Others, like Márcio, saw no value in this approach. Rodrigo had to find a way to navigate between them, and his new approach of relying on intrinsic motivation was making this task more challenging.

We returned to work. I read a section Rodrigo had drafted, then proceeded to work on the next section, which we had agreed I would write. Rodrigo, meanwhile, dedicated himself to implementing a change to the Kepler code that we had discovered we needed to present a more elegant example in our chapter. (The chapter was describing a version of Kepler that had yet to be released, so we had the freedom to change Kepler to fit our description of it.)

By 4:30 p.m. Rodrigo had completed the change and sent out a lengthy email to the Kepler mailing list, describing the problem we had encountered and the proposed solution. It concluded with a request for comments: "What do you think about the change?" As he finished the message, he shouted out, as if talking to the people on the list: "Answer! Because I am lonely here, sitting in this room with a crazy Russian guy!" Indeed, Rodrigo's pursuit of his open source vision was straining his local ties, and if the hypothetical foreign contributors were to remain silent, he would be left with little company other than his resident ethnographer. After reading Rodrigo's message, I posted a brief response, pointing out an additional benefit of the proposal and raising the possibility of an additional change. Rodrigo saw the new message in his inbox. "Oh, someone answered!" he exclaimed. "Oh, my friend Yuri!"

As I read Rodrigo's message, however, I noticed a problematic ramification of the change we had proposed, realizing that it would break something else that our examples depended on. I turned around and mentioned this to Rodrigo. We went to Nas Nuvens's lobby, planted ourselves in the two beanbags, and proceeded to discuss the problem for a long time, eventually coming up with a better solution. Rodrigo returned to his seat and wrote another email, describing the problem we had found and the new solution. "Again, if someone is still with me, comments are welcome," he concluded.

Another hour later, while Rodrigo and I were having a late dinner around the block (a different place from where we had lunch, since we were celebrating the day's success), a list member "John" responded to Rodrigo's second message with a short follow-up question. His message ended up being the only one in this thread written by someone outside Rio de Janeiro. (Of course, we did not know where John was, but it seemed safe to assume he lived abroad.) Rodrigo responded to John's question after getting home; I replied to Rodrigo from home in the middle of the night, this time disagreeing with his response. Another message from Rodrigo, just a few minutes after mine, was the last one in the thread. Next morning Rodrigo committed the change to the code repository.

This email conversation on an English mailing list, in which all but two sentences were written by the two of us, who spent most of our day sitting in the same small room, could seem strange and almost farcical. We were resolved, however, to continue developing Kepler in an "open" way, hoping that someone would join us eventually. Turning Kepler into a global project, a translocal place where space would seem to no longer matter (and where foreign contributors' labor would be marshaled from Rio de Janeiro), would require substantial work on the ground. We did not expect it to be easy, and even small successes counted.

As we were having dinner, I asked Rodrigo if he really felt that he was alone, talking to himself, as some of his earlier comments suggested. No, he said. He had a hundred people on that mailing list, after all. He did not expect them to write code, it was enough that they read what he wrote and commented, if only occasionally. This, pointed out Rodrigo, was *much* more than what they did up until March! I asked Rodrigo if his hopes were really limited to the foreign participants just asking questions rather than actually making contributions. He did not *expect* them to contribute code, said Rodrigo. But maybe they will, he added.

Megalomania

Rodrigo's distinction between his expectations and what *might* happen reminded me of another conversation from earlier that day. In the early afternoon I had stepped outside of Rodrigo's office to get some water. There I ran into Pedro, one of the Nas Nuvens's employees whom we met in chapter 2. Pedro worked in customer support as well as software development. While he was still working on his undergraduate degree, and was doing it through a night program at Estácio de Sá, a private university that could hardly be compared to PUC, Pedro was considered by Rodrigo to be one of the most promising people at Nas Nuvens. (All PUC graduates had long left for better salaries. A few months later Rodrigo helped Pedro get into PUC's prestigious master's program in Computer Science. Another year later Pedro left the company, having become overqualified for the salary Nas Nuvens could pay him.)

Pedro was surprisingly dressed up: black pants, a white shirt, and a tie. He had sunk himself into a beanbag, which made his attire seem even more out of place. It turned out that he was presenting the final project for his undergraduate degree later that day—a web application written in Ruby-on-Rails, a recently popular web framework for Ruby, a programming language somewhat similar to Lua and seen by some as one of its main competitors. (Developed in Japan, Ruby was little known for years until it suddenly exploded in popularity. As another *peripheral* language and a former underdog that had at last made it big, Ruby had a special place in the imagination of the Lua community.) We talked about the report Pedro had to turn in for the project. I asked in which language he wrote it. "In Portuguese," he replied. "It *has* to be in Portuguese." Intrigued by his "has to be," I asked in what language he had written the code. The code was all in English, explained Pedro. Unlike the report, the code *could* be in English, since only the advisor had to look at it, and the advisor knew English. So it *was* all in English, he continued: the names of functions and variables, as well as the comments—all but the report.

"Why?" I asked. Pedro laughed before answering. "Out of megalomania," he said then. He explained that he and his partner wanted to think that one day people abroad would be using their code, perhaps even contributing. Writing the code in Portuguese would exclude all of those people. Of course, making their code in English excluded some Brazilians too, continued Pedro. But those people were already excluded. The code was based on Ruby-on-Rails, which was documented only in English. Ruby-on-Rails function names were in English too. One could not work with it without knowing English. If Pedro were to start a company around his final project, he would not consider hiring anyone who did not know English. But again, he summed it up, it was also about megalomania. What if he decided to hire a developer in India? If his code were in Portuguese, he would not be able to do this.

Was Pedro joking? On the one hand, he had to be. Someone in California talking about "hiring a developer in India" had most likely worked closely with developers from India (perhaps having found himself on a few occasions as the only non-Indian in a conference room) and would likely know people who had traveled to India's outsourcing capitals. The decision to hire a developer in India would thus be a practical question, a matter of cost-and-benefit analysis. Things could not be more different for Pedro, who had never met anyone from India or even anyone who had traveled there. Pedro's wages (and those of most programmers in Rio) were hardly high enough to justify hiring programmers in India to save money on salaries. The talk of outsourcing was thus not a matter of planning for cost savings, but a matter of dreaming about one day entering the same league as the big global players.

This dream was so far-fetched that Pedro himself called it "megaloma-nia." Having learned those global dreams from the global technical culture in which he engaged virtually, Pedro maintained an ambiguous attitude toward them. This was not something he was ready to defend publicly as a plan for action, and he was ready to laugh at his own global imagination. Yet, he did write all his software in a foreign language, entertaining the pos-sibility that those dreams *just might* come true.

Pedro's story made explicit the ambiguity of global imagination that came up in more subtle ways in numerous conversations with Brazilian software practitioners, including those involving Lua. This ambiguity could be understood in two ways. One approach is to liken it to what Favret-Saada (1980) calls "I know . . . but still . . ." in her discussion of witchcraft in France (51). Favret-Saada describes the ambiguous approach to witchcraft, which combines the public acceptance of the rational view with a sup-pressed belief in witchcraft, rarely verbalized and quickly withdrawn upon questioning, yet strongly affecting what people do.[5] Pedro's plans show a similar duality, being powerful enough to actually affect practice, yet not defensible in public and eagerly labeled "megalomania" upon interroga-tion. A somewhat different, but closely related, approach is to look at such global dreams as a game of make-believe, a simulation of a global practice. Regardless of whether Pedro's project had a global future, writing the code in English and imagining future plans for hiring programmers in Bangalore could similarly be *entertaining,* a way of doing in imagination what could not be done in reality.

Regardless of the attributed motivation, however, it is important to rec-ognize the commonality and the powerful effects of this form of imagina-tion, which I call "subvocal" (by analogy with "subvocal speech"—a form of speech that involves actual movement of muscles without producing audible words). While dismissing his plans as "megalomania," Pedro did write his code in English. If he were to start a company and hire other Bra-zilian developers, such developers would have to work with Pedro's English code. Pedro's subvocal imagination was therefore having tangible effect on his own practice and possibly that of others.

Like Pedro, Rodrigo had a "megalomaniac" dream of running from Rio de Janeiro a major international open source project, a web develop-ment platform that could compete with the popular frameworks for Ruby, Python, and Java. Like Pedro, he repeatedly described his dream as "crazy," a part of his half-humorous plan of "world domination." The word "crazy" (*maluco*), was in fact one of the most common words that he applied to himself, his work, and the people he respected. Unlike Pedro, Rodrigo had actually invested a number of years of his life into his "crazy" dream *publicly*. He thus could not as easily dismiss it as a joke. Instead, he embraced the image of a crazy dreamer. Even so, many of his plans fit within the "I know . . . but still . . ." space.

The Windows Build

As we celebrated the day's success that Wednesday, Rodrigo was in a positive mood, focusing on what Kepler had achieved rather than on the project's troubles. This contrasted sharply with many of the days from the previous month, during which the themes of "being wrong," being "alone," and just "giving up" had come up again and again in our conversations. (The difference, I suspected, had much to do with the fact that the two of us were now actively working together side by side.) Two weeks earlier Rodrigo had shown me a checklist page he had created on a wiki. The problem, he said, was that nobody seemed to want to follow it. Tiago and others did not even think Kepler should be making releases. Perhaps they were right and he was wrong. He had fought with the Lua community for a very long time, trying to make Lua into something that the majority of the users seemed to have little interest in. Perhaps he was just wrong. He was all by himself— even those who were working on Kepler with him were not in agreement. (He was not *planning* to give up, Rodrigo explained on a number of occasions, but the possibility was always there on the table. After all, while Nas Nuvens could not give up on Kepler, *he* could do just that, perhaps joining his friend Renato as a manager of Java developers.)

One particular source of contention was the installation process. Like other software systems with a part written in C, Kepler's code had to be compiled or "built" before it could be used. This process—"the build"— could in theory be left to the users, and the Lua community had historically stressed this approach, since most users of Lua were themselves skilled developers working on software products written in C and were assumed to be capable of handling the compilation step. It meant, however, that compared to other languages used for web development, Lua was quite difficult to install. This created a problem for Nas Nuvens, whose customers did not have the expertise to compile software on their own. Simplifying the installation procedure was therefore one of Rodrigo's main projects in the spring of 2007.

The problem could be broken in two: one approach for users of Unix-based systems such as Linux, and another for users of Windows. The first problem was easier: users of Unix could be expected to have the software tools needed for software compilation and be accustomed to compiling source code into executable software on their own machines, at least as long as the process was automated and did not require too much tweaking. Rodrigo also had access to Unix enthusiasts among PUC students. In 2006, he used FINEP money to hire Alan, then a PUC master's student. By April 2007 Alan had written a robust "build script"—a program that could be used to compile and install Kepler on a wide range of Unix machines, which I then helped test and document. With the new script, a user could download, build, and install Kepler in less than a minute, making the process of installing Kepler on Unix computers quite trivial. Among other things, this made it possible to easily set up Kepler on Dreamhost, one of the most popular solutions for cheap web hosting at the time. By mid-April we were even "self-hosting": Kepler's new site was driven by a rudimentary wiki engine that I had implemented in Lua using Kepler's software. Since almost all Kepler contributors used Linux or other variations of Unix for their own server needs, the general opinion seemed to be that Kepler 1.1 was done.

Rodrigo, however, wanted to have a version of Kepler that could be used on Windows—partly due to the fact that he believed this would open a wider "market," but mainly because nearly all of Nas Nuvens's costumers ran Windows. (Rodrigo and many others sometimes attributed this to their being in Brazil, not yet on the Linux bandwagon.) Making Kepler work on Windows was a challenging task, however, both because of the quirks in the Windows build system and because of the Windows users' high expectations for the ease of the installation process.

A further challenge lay in finding people to do the work. The developers to whom Rodrigo had access appeared to be divided into those who did not have the skills for the task (e.g., the developers employed by Nas Nuvens) and those who had the skills but lacked interest. Tiago, in particular, had the knowledge necessary to solve this problem (and had done this for the earlier release), but was more interested in other aspects of the project. Following the new policy of relying on developers' intrinsic motivation rather than simply just hiring them to do tasks, Rodrigo decided to not press, hoping instead to find volunteers among the list subscribers. A few people expressed interest in helping (a big improvement from February, noted Rodrigo), but none were willing to lead the task. By early May Rodrigo had to face the fact that, if there were a Windows build, it would have to be done by *him*.

Rodrigo's attempt to do the Windows build, however, made it clear that despite good high-level understanding of the technology behind Kepler, he was unable to make sense of the details of the software, which had been written by others. More generally, he was out of practice even when it came to Lua programming, not to mention compiling C code on Windows. Rodrigo struggled with the build scripts, occasionally having to rely on *me* to help him out. As frustrating as this process was, however, the Windows build proved a blessing in disguise. As Rodrigo was starting to understand, getting his programming skills back was not just a matter of solving the specific practical task of completing the Windows build. It was also a matter of adjusting to the change in the rules of the game brought about by "opening" Kepler, following through on the course he had already chosen.

As many people pointed out to me, in Brazil, where less educated technical workers are abundant and cheap and where many needs of the domestic market do not require high skill, highly educated people like Rodrigo get promoted quickly out of programming jobs into management. Those who are thus promoted often lose their hands-on programming skill quickly, in part due to the fast pace of change in the software technology. They often lament the loss, but stay in management. Rodrigo's friend Renato, working as a manager in a local software company, often talked about his desire to get back to programming, carrying a copy of *Programming in Lua* in the back seat of his car. Renato's management job, however, kept him quite busy and left him exhausted at the end of the day. Renato still wanted to think of himself as a "computer scientist," but appeared to be "stuck" in management work indefinitely.

Rodrigo, whose first job after leaving the university similarly involved managing software developers rather than writing code, had stayed more closely involved with technology, but had similarly become more skilled at bringing together people, ideas, and resources than at getting code to compile. Rodrigo usually described his role in the project as "an architect"—a term that denotes someone who provides the larger technical vision for the system. At times, however, he suspected that his role was turning into that of "a PHB"—the "pointy-haired boss" from Dilbert cartoons.

Rodrigo's desire to develop (or, rather, lead the development of) something more than a customized system for a local client, required software developers with a higher level of skill, who would need to be paid high salaries. If Nas Nuvens had been successful in attracting the additional investment, Rodrigo could proceed to hire such programmers and could himself focus on directing the work. The small and unreliable funding that the project was getting, however, was proving to be insufficient for this. "As far as C programmers go, I know a few," said Rodrigo, "As far as C programmers who want to work for free on a Lua platform—I don't know any."

The dearth of local programmers willing to work for free (or for unreliable pay) on a Lua platform could be solved in two ways. One was Rodrigo's local "culture farming" efforts, getting PUC graduates interested in the project through a combination of financial and cultural rewards. As Rodrigo had decided earlier that year, however, the main solution had to involve attracting foreign open source developers. He soon discovered, though, that getting the help of external developers required an entirely different currency, one he had in even shorter supply than money: respect earned through demonstrated technical competence. As Rodrigo explained to me in an interview:

Rodrigo: I noticed that on a free [software] project, there are two currencies of exchange: credit and respect. And there is a tangible good, which is the *source* [i.e., source code, says in English]. So, you have two abstract currencies, and one concrete. I can provide *source* to the community and with this earn respect. Or someone can provide me *source* and I could pay with credit. [. . .] If I give someone credit, they get respect. And what I am trying now is to earn respect.

While Rodrigo's words suggest the possibility of a simple conversion of code into respect into more code, Rodrigo understood that the free software community assigned respect primarily to the people who wrote the code, not to those who organized and funded its production.

Rodrigo thus found himself in a paradoxical situation. "I was in a very delicate situation," he explained, "because I had a platform that I had come up with—the idea of the platform was mine—but that I didn't know how to use." His approach to the project had been to identify the pieces he needed, find money for them, and then find developers capable of writing them. While this all required substantial organizational work, from a *technical* point of view the pieces just appeared. "I never stopped to look," he explained. "I said: 'I need a LuaExpat!' And it appeared. Ah, good, now I have this piece. 'Now I need a LuaZip!' It appeared. 'Now I need MD5.' It appeared."

Based on what Rodrigo had read about open source—for instance, Raymond's *The Cathedral and the Bazaar*, which he that he had read back when it was still an online article and then reread when it came out as a book (Raymond 1999)—Rodrigo knew that open source projects were supposed to proceed differently, starting from an individual programmer's desire to "scratch a personal itch," that is, solve a specific problem through actual programming work. They were not supposed to proceed top-down as a government-funded pursuit of a grand technological vision. Rodrigo's project, as he himself saw it, was proceeding "backward," having started in the wrong place.[6]

Running the project by the local rules did not require Rodrigo to understand the details of the platform that was being built under his leadership. Local participants had accepted Rodrigo's role as the provider of an engaging vision and the resources they themselves could not obtain. Such resources, however, had no power in an international free software community to be supported by volunteer efforts. Rodrigo had to obtain this new currency. For this he had to learn to engage in his own project in a new way: as a developer capable of discussing the minute details of the code with the members of the list and of making changes to this code when necessary. The Windows build of Kepler was forcing him to do this work. In our interview a few months later, Rodrigo described the time as a difficult but important experience, which involved deconstructing his mental model of how the project worked and of what his place was in it, then replacing it with a new one. Yet more important, it involved putting the new model into practice.

The New Dynamic

I left Rio in early August 2007, but continued to follow the project remotely, partly out of a researcher's desire to know what happened later, but also as a simple matter of commitment. In April I had volunteered to write a wiki based on Kepler. The wiki became the first public application built on top of Kepler, and quickly came to be seen as a demo of Kepler's capabilities and proof that Kepler could actually be used to build real web applications.[7] I now had a role in the project that could not be easily dropped.

By September, I had started to notice a substantial growth in list traffic. The list received 236 messages in August and seemed headed for setting another record in September. I was finding myself barely able to keep up with the list. I called Rodrigo to catch up on what had happened since my departure. Rodrigo started the conversation by communicating his excitement over the recent changes. The opening of Kepler was finally bringing results. The activity on the list was an important part of that: Rodrigo had also noticed the clear spike in traffic. But it was not just the number of messages, he stressed, but the changing dynamic. The list members were no longer just asking questions about Kepler: they were making contributions, and those contributions tied together Kepler's global vision and Nas Nuvens's local problems.

Rodrigo offered an example of collaboration that illustrated the list's new dynamic. Two weeks before our interview, while working on one of Nas Nuvens's projects (which were starting to use Kepler), Pedro discovered a particular problem specific to Windows. He sent a message to the Kepler list, which went unanswered. After spending some time investigating the problem on his own and exchanging messages with Tiago, Pedro managed to identify two potential causes of the problem. He sent a new message to the list, now much more specific, and two foreign members of the list joined in the discussion—one of them from Uruguay and another from Southern California. In the course of a lengthy discussion that involved Rodrigo, Pedro, Alan, and the two foreign members, the California engineer proposed using a library developed by Luiz Henrique, one of the authors of Lua. The library did not work with Windows, which led to an additional conversation, off the list, with Luiz Henrique. In the end, the California engineer adapted the library for Windows. Rodrigo and Tiago then jointly made the necessary changes to Kepler.

This intense collaboration, which brought together Rodrigo, a Nas Nuvens employee, FINEP-funded Kepler contributors, a member of the Lua team, and two foreign participants served as an example of the project's new dynamic. It helped Nas Nuvens solve a specific problem it faced while resolving a serious underlying problem in Kepler and improving the quality of the platform on Windows. The increasing ease of using Kepler on Windows led to increased growth in interest among local companies. In particular, João was now talking to local firms and FINEP about the possibility of bringing Kepler into digital TV projects.

The project was also helped by the arrival of Jason, a Rio developer we met in chapter 3. In the spring of 2007 I noticed Jason's name on the Lua mailing list, as one of a few names that sounded Brazilian yet had not appeared on my map of the local Lua community built around PUC. I emailed Jason and discovered that he was working for a local company in Rio. We scheduled an interview. While a substantial part of the interview dealt with Jason's early steps into the software profession, which I described in chapter 3, we also talked about Jason's use of Lua. As Jason told me, he had discovered Lua several years earlier, while working on a computer game as a hobby project. He had abandoned the project but later remembered Lua when he faced a problem at work for which it seemed like a perfect solution. Having used some of the modules maintained by Kepler, Jason also talked with much interest about this project and his own desire to participate in something like this. Noticing that Jason seemed to have the exact combination of expertise and interests that Rodrigo was desperately seeking, I asked him if he had thought practically about participating in Kepler and suggested that he should meet with Rodrigo.

Despite his excitement about the fact that Lua was developed locally, Jason had never approached any members of the local Lua community in person. Despite living in the same city, Lua's authors seemed too far removed from him.

Jason: So, when I discovered that the guys were at PUC, that I started using something and the guys were from PUC, I was like: "Damn. This is there. I think I'll go there and see what it's like." But I wouldn't even know were to meet those guys. I think if there were a course they were offering, I would go immediately, running. But I wouldn't go and knock on the door to learn where the guy is.

There was no such course, however, as the Lua team seemed, at the moment, to have little interest in reaching out to local developers such as Jason. The authors of Lua's modules, such as Tiago or Márcio, also seemed quite distant.

After returning from our interview, I sent Jason an email encouraging him to talk to Rodrigo. A day later they met for lunch. A few weeks later Jason was actively involved in Kepler. For Jason, meeting Rodrigo was a transforming experience. As he told me in a later interview, Rodrigo showed him that one could live in Brazil doing something outside Java and Microsoft's .Net. While he had always wanted to work on something like this, Jason explained, he never thought it would be possible, as the bills had to be paid. "We don't just live by the economy," he told me, "but speaking like modern thinkers, it's a factor that cannot be ignored. [. . .] I've always wanted to do this, but how would I make a living? If I were to be doing those highly experimental and advanced things, which do not have . . . are not commercial in a mainstream way? How would I live outside Java and .Net?" Rodrigo was a living proof that this was possible. "Well, he is there and living, right?" said Jason.

Jason's entry into the project, however, also proved a lifesaver for Rodrigo. While a number of new subprojects were generating a lot of interest, several others were stagnating, in part due to the fact that Alta's Fabio and Fernando were increasingly busy with Alta's expanding projects and had less and less time for Kepler. Rodrigo's reliance on FINEP funding, however, meant that projects could not be simply dropped for lack of interest. Jason's adoption of one such lagging project helped Rodrigo focus on the other projects, the ones he felt were more promising. Another month later, Jason left his job and came to Nas Nuvens, taking over Rodrigo's de facto role as the company's technical director. This further freed Rodrigo to focus on Kepler: he was now able to spend as much as 80 percent of his time working "as a developer." A month later, Rodrigo managed to dedicate some of this time to do a side project, for which he wrote half of the code. This new role was making him more comfortable in his position as a technical leader of an open source project.

As I was spending my days transforming my field notes into early drafts of chapters (and spending my nights adding new features to my Kepler-based wiki software), Rodrigo appeared to have achieved exactly what he aimed for in March: an assembly of local and global resources that was allowing him to move ahead with the project while also solving Nas Nuvens's local problems. The project was also an increasingly transformative experience for developers such as Jason, who now had a chance to participate locally in the construction of a global platform.

The new dynamic continued for several months. At the year's end, however, the project found itself without money again. A grant had been awarded for the new year, but no money had arrived, having disappeared somewhere in the complicated network of funding transactions. The work, of course, was expected to proceed on schedule. Over the next many months Rodrigo had to suspend his newly acquired career as a developer, dedicating much of his time to the work that he thought he was starting to put aside: finding out what had happened to the money the project was owed and what would need to be done to get it back. The team's morale was also seriously hurt, as some of the contributors had to go for months without getting paid for their work. The team managed to release the next version of Kepler in June 2008, but after that the project was effectively suspended. When I returned to Brazil in December of that year, Rodrigo had managed to finally get access to the money awarded a year earlier and was making efforts to bring the project back to life, but a lot of momentum had been lost and a lot of work had to be redone.

As I was completing my UC Berkeley dissertation in early 2009, I shared my drafts with Rodrigo and other people I had met in Rio. After reading the draft, Rodrigo commented that my exposition had shown to him what he had already suspected: the futility of his project. What he had tried to do was simply not possible in Brazil, he said. He noted that seeing the phrase "a decade of his life" in my chapter made it particularly clear to him that it was time to move on. By the time I submitted my dissertation, Rodrigo had quit his job at Nas Nuvens and started looking for other projects. Even before his departure, Nas Nuvens had started a transition from Kepler to Drupal, a popular open source content management system written in PHP.

Rodrigo's departure did not quite spell the end of Kepler. In a way, it just completed the transition that Rodrigo had started when I arrived in Rio in March 2007. Even in 2008, while the work on the project slowed down substantially following the loss of funding, it did not stop entirely. Rodrigo appeared to have succeeded in convincing the contributors to see the components they were working on as their own projects, with Kepler's serving primarily as an umbrella project helping to coordinate the interaction between the subprojects and attract financial support. Rodrigo's departure just finalized this new arrangement. Tiago continues to maintain the key components of Kepler's web server. He does so in loose collaboration with an American programmer who lives in Argentina, and with occasional contributions from others—for example, a Russian developer working for a company that uses Kepler in Moscow. Alan's build script, originally developed for building Kepler, has grown into a more general solution for automatic installation of Lua components, which today covers nearly two hundred libraries and is recognized as one of the main ways of installing Lua modules. Some of the modules developed by Kepler have been taken over by Lua developers outside Brazil. The traffic on *lua-l* suggests that many of Kepler's modules remain in use. Most of them can today also be installed with a single command on Ubuntu Linux, thanks to the efforts of a developer in Italy. Kepler's software is also starting to face competition from alternative solutions for Lua web programming. While such alternative solutions might point to shortcomings in Kepler's technical design, in a way they bring closer to reality Rodrigo's broader vision of Lua as a platform for web development.

What is missing, perhaps, is any clear sign that Lua is being used much for web development in Brazil.

* * *

This chapter has looked at a small project that brings into focus the complexity of globalization of modern technical work. Unlike the cases of Alta and Lua that I discussed in the earlier chapters, Kepler's case does not easily allow for simple analysis. It does, however, show the application of many of the ideas of the framework that I presented in chapter 1.

Kepler's story highlights the role of imagination that I stressed in many earlier chapters. The project proceeded from a "crazy" dream—much like many other dreams that we had encountered before—for example, Marshall Montenegro's plan to build airplanes in Brazil. (Montenegro's plan, of course, became successful beyond anyone's imagination: Brazilian airplane maker Embraer, based in São Jose dos Campos, is today one of Brazil's largest exporters.) Carrying out this dream required assembling alliances. Doing so was made easier by the fact that other local parties were pursuing their own globalization projects, for which Rodrigo's project could serve as an ally. Rodrigo's brother was seeking to build a Silicon Valley–style technology company, and Rodrigo's project could provide him with the requisite "local innovation." FINEP was looking for innovative approaches to funding innovation, and in particular eager to try its hand in promoting open source, as some of the foreign governments had done before. As is common at the periphery, however, the allies must constantly reevaluate their allegiances, considering whether they would be better off building direct ties with the larger world. For Nas Nuvens, for example, the emergence of globally popular platforms such as Drupal posed the question of whether it was time to cut losses and switch to those new technologies. For Rodrigo, it meant looking for ways to reduce his dependence on Nas Nuvens and local developers and figure out how to enroll foreign labor into his project. This meant finding a way to "relocate" the project from the offices in Rio de Janeiro into the virtual space created by the global network of Lua developers and mediated by a technical infrastructure composed of tools such as mailing lists and wikis. It also meant looking for ways to move the project away from its financial foundation in Brazil, relying instead on a system of labor managed through flows of cultural currencies: credit, prestige, and fun. Ironically, this also meant reducing the projects' dependence on Rodrigo's original role in it: the manager or, to use Rodrigo's term again, "the PHB." Rodrigo attempted to adapt his role in the project, becoming a developer actually writing code. In the end, however, he concluded it made more sense for him to move on, leaving the code in the hands of younger programmers like Alan and Tiago.

9 Conclusion

This book has looked at the world of software development from a somewhat unlikely place—Rio de Janeiro, a city widely known for its beaches and music, but rarely for its software. Looking at software from such a place, however, provides us with a useful perspective on globalization—of software, of technical practice, and of skilled work more generally. It highlights a seeming contradiction in our thinking about globalization: software development is often described as an immaterial and placeless line of work, yet it is dominated, both economically and culturally, by a small number of places. This contradiction appears not only in scholars' accounts of software work, but in developers' own accounts. "A server is a server," Rio developers say, highlighting the similarity of their work to that of their California colleagues. Yet, at other times, "this is not Silicon Valley" comes up as a frequent explanation. This paradox is hardly specific to Rio or Brazil. After all, despite Silicon Valley's yet-to-be-challenged pre-eminence in the world of software, the overwhelming majority of software developers live in places that can also be aptly described as "not Silicon Valley." And even though the domain of software today brings this puzzle forward in the clearest way, the underlying contradiction is hardly specific to software.

Understanding this contradiction required looking closely at the work of Rio developers. It entailed first of all recognizing the many ways in which place continues to matter in today's globalizing world even in a supposedly "global" field such as software development. Throughout the book we have seen numerous examples of how software developers' work depends on local networks, local relations of production, local institutions. We have also seen, however, the many ways in which Rio developers' work is in fact quite similar to that of software developers elsewhere, and the many ways in which they are connected to remote places. Most important, however, we saw the *making* of such ties.

I have therefore tried to show not only how being in a "wrong place" makes software work more difficult, but also how peripheral practitioners work to overcome such disadvantages—and how their daily (and often unnoticed) work helps software technology acquire its seeming universality. In other words, I have attempted to show the importance of recognizing peripheral participants as neither happily "connected" to their remote colleagues nor as woefully "disconnected" from them, but rather as actively working to build connections to remote places and re-create locally a remote practice. I showed how such attempts sometimes fail and sometimes succeed. Their success and failure often depends on assembling configurations of local and remote resources and on the continuous renegotiation of such configurations. Despite the many difficulties and setbacks experienced by the peripheral actors, their projects over time bring about an increasing synchronization between the local and remote contexts, which in turn facilitates further synchronization of the practice.

This process is arduous and slow. It is far from complete today. It may in fact never be complete, due to continuous changes in the practice in remote centers. And, as I also tried to show, peripheral developers' efforts often serve to give additional power to remote centers. This process also often leaves the participants in a paradoxical state of being hyperconnected in some ways and quite disconnected in others. It sometimes creates seemingly bizarre configurations that involve local participants connecting via remote centers—as, for example, illustrated by Luciano's learning English in order to read a book written by a fellow Portuguese speaker about a programming language developed just a few kilometers away. It is this process, however, that ultimately gives rise to what we call globalization.

Software as a Global World of Practice

To make sense of the seeming contradictions in the experience of Rio's software developers, I looked at the system of activities related to software production as "the world of software"—a case of a *world of practice*, the idea developed theoretically in chapter 1. The notion of "a world of practice" provides an important theoretical counterweight to the idea of "place." Armed with this concept, we can look at Rio developers as simultaneously engaged in two contexts: the local place and the world of software. This allows us in turn to recognize the existence of a shared global context that unites software developers around the globe, while at the same time asking how this global context is created and maintained, and how it relates to specific places. In what follows I highlight some specific aspects of worlds of practice and the world of software in particular, noting how they were illustrated in the book.

Culture and Economics

I looked at the world of software as being *tied simultaneously by cultural and economic relations*. I tried to show throughout the book that the work of software developers cannot be understood without considering, on the one hand, the extent to which it is affected by the *culture* of software development, namely, by the shared dispositions and techniques, acquired through active engagement with communities of practitioners, and the cultural rewards involved in being recognized (and recognizing oneself) as a legitimate member of the world of software. As we saw throughout the book, what software developers do is affected strongly by what they see as "cool," "fun," or "elegant" versus what they see as "boring" or "ugly." This idea was discussed most explicitly in chapter 3, where I looked at the developers' early steps into the world of software, but it was then exemplified repeatedly in subsequent chapters. My discussion of Kepler, Lua, and even Alta showed the participants' interest in the cultural rewards provided by their work.

On the other hand, it would be wrong to ignore the fact that in most cases software work is done in the context of employment—a politico-economic relationship in which the developers offer their labor in exchange for resources that they can use both for basic sustenance and the acquisition of objects required to participate in the cultural side of the practice. (This would include, for example, the acquisition of the latest gadgets, or using money to hire others to assist in the pursuit of a culturally motivated technological vision.) The material side of software work was discussed most explicitly in chapter 5 and at the end of chapter 3, but was exemplified throughout the other chapters as well. Even when it was temporarily put in the background—for example, in parts of my discussion of Lua—I tried to remind the reader that such backgrounding of economic concerns is only possible because of a particular arrangement of labor relations—for example, the Lua team's privileged position within the funded academic research system. I also tried to show that material concerns remain just as important in open source software development, even as this mode of software production may put a heavier weight on cultural rather than economic means of control over labor.

While cultural and economic perspectives on work are both common, they are rarely combined. This book shows the need to do so, by demonstrating how neither of the two is sufficient by itself. A purely cultural perspective (which often seems to dominate, for example, the discussion of open source software development) would lead us to politico-economic naiveté and unjustified expectations about the upcoming "flattening" of the world. (It also can lead us to accepting too quickly the assumptions of the culture we are studying, as we would have no basis from which to critique it. This again is common in the literature on open source.) On the other hand, purely economic approaches to work, and the focus on control of the labor process, would lead us to put too much stress on formal organizational systems, such as firms and industries. While such entities do play an important role, I believe it is important to recognize that knowledge and innovation in software is often (and increasingly) produced and shared through lateral ties between individual developers, who are often driven as much (or more) by cultural motivations as by economic ones.

Combining the two perspectives means, among other things, looking at how the developers themselves reconcile the cultural and economic sides of the software practice, a task that is rarely easy. "Why will no one ever pay you to do anything interesting?" asks a message to the Lua mailing list. The question is asked in jest—many software developers stress that being paid to do what is interesting is the biggest appeal of software development as a profession. (Needless to say, this means doing what *software developers* find interesting, since the desire to spend long hours "mapping interrupts" is hardly a universal human trait.) It highlights, however, the frequent challenge of simultaneously extracting cultural and economic benefits from one's work. I tried to show the interaction between such factors in several chapters, from Jason's stories in chapter 3 to Rodrigo's work on Kepler in chapter 8.

Reproduction of Practice in a New Place

My discussion of software development in Rio de Janeiro positions the city as a peripheral site in a widely dispersed but highly centralized world of software development practice (chapter 4), which is dominated by a small number of "meccas." Local participants orient themselves toward such meccas in an attempt to draw on their symbolic power and to bring the local practice closer to the remote standards. At many times during my fieldwork, I found myself in a privileged position as a visitor "from the Valley," often given credit for knowledge that I did not actually have or recruited to serve as an arbiter of local value.

Understanding the distributed-yet-centralized nature of worlds of practice requires paying attention to the process by which a system of activities that originates in one place is later reproduced in other places. The practice of software development in Brazil must be seen as *a partial and ongoing replication* of the practice of software development based largely in the United States. I stress the role of individual participants in this process: the replication takes place as many individual globalization projects, each driven by someone's desire to engage locally in a remote practice, in pursuit of either cultural or material rewards. This leads me to highlight the "diasporic" situation of the peripheral practitioners, who engage simultaneously in two cultures: the local mainstream culture and the foreign culture of the practice, illustrated, for example, by the analysis of developers' use of English and Portuguese.

My discussion of the process of reproduction draws on ideas of *disembedding* and *reembedding* (Giddens 1991). A practice, understood as a system, cannot move to a new place all at once. Individual elements of the practice, however, can be detached from the system ("disembedded"), moved and inserted ("reembedded") into a new context. Such mobile elements may include material objects (the UNIVAC brought to Brazil in 1960 or today's mobile gadgets), people ("the Wallauscheks" or even myself), ideas ("the Smith Plan" imported from the United States, or Rodrigo's ideas about open source software developers), documents (the different books read by Rodrigo and others). As we saw, people attempting to engage in the practice in a new place must reassemble it from disjointed elements brought from other places, and such reembedding is often a nontrivial task.

The same applies in reverse: peripheral participants who want to make a contribution to central practices must thoroughly disembed their innovations, making them mobile. As we saw in the case of Lua and Kepler, such disembedding does not involve conversion of contextualized elements into some neutral and context-free medium. Rather, it involves loosening them from the local context and linking them to the global context of the practice, which, however, is often *local* for those in central sites. Knowledge once shared through Portuguese conversation, for example, takes the form of a global book, written not in some neutral Esperanto or Volapük, but in English, the language spoken fluently in California but significantly less so in Rio de Janeiro. The price of such disembedding is borne not only by the peripheral innovators who must undertake the work of disembedding, but also by peripheral users. Luciano's struggle (in chapter 2) with reading *Programming in Lua*—an English book written in Brazil—is indicative of this. It also shows how peripheral participants in many ways bear the burden of maintaining the predominance of central sites. Such disembedding can also draw new boundaries locally.

Building on the idea of practice as a system, I stress *the cumulative nature of the reproduction process*. The process of reproduction of practice across space happens over time, as a gradual synchronization of context. At each step, elements are brought together and local work is done to make the context more similar to the central sites, thus laying the "tracks" that Latour (1987) stresses must be in place for knowledge to move between places. Alternatively, we can think of such efforts as creating landing strips for future elements—enclaves of the practice in the midst of otherwise unconquered territory, much like the actual landing strips that Marshall Montenegro had to build throughout Brazil as he worked to establish aviation in the country. Once the tracks and landing strips are there, importing additional elements becomes easier. Chapter 4 shows us how over a number of decades Brazil's context was brought closer to that of the central sites of the computing world. Establishment of connections to the Internet, for example, transformed the methods for keeping practices in sync. The different projects described in the last four chapters all contribute to the continued synchronization of context.

Seeing software practice as itself an element of a larger system of practices has led me to stress *the parallel nature of the reproduction process*. As we saw most clearly in chapter 4, the reproduction of the foreign software practice cannot be understood in isolation from the parallel efforts of people engaged in *other* practices, all of them pursuing their own globalization projects. The fact that the centers of many practices coincide simplifies this task tremendously. The work of Alta's engineers, who have mastered technology developed on the West Coast of the United States, is made much easier by the fact that their clients seek to emulate business practices originally developed in the same place. This parallel reproduction, however, also raises the stakes. The clients' concerns about the successes of their own globalization project lead them to seek *authentic* practitioners of supporting practices. To the extent that they can afford it, they want software developers who can provide them with the best of the *world's* technology, not just with *local* substitutes. This in turn means that the developers are expected to project a global image in everything they do (both for the clients and for each other) and highlight their ability to transcend space in their practice. "A server is a server" thus becomes not just a statement about the state of the world, but a promise that developers must fulfill.

Reflexivity, Imagination, and Collective Action

The parallel nature of the reproduction process leads to *a complex relation between individual and collective efforts of reproducing foreign practices*. The local practitioners must often make a decision whether to cast their lot with local colleagues or to focus on their individual connections to remote centers. We saw examples of this both within and between practices. For example, the Brazilian government and other Brazilian users of computers had to decide at several points whether to rely on local makes of computers (in the hope of eventually benefiting from cheaper technology more attuned to local needs) or to focus on acquisition of the better foreign machines. Within the practice, software developers in Rio de Janeiro must decide whether to go with the globally established programming language such as Java, or to dedicate their efforts to supporting a local one. I use the cases of Lua and Kepler to show the challenges this presents for local innovation, which must often succeed abroad before being accepted at home.

To understand the unfolding of such collective projects, we must pay attention to reflexivity. We saw throughout the book that peripheral software developers know quite a bit not only about the social context they inhabit (as do most people, argues Giddens [1979]), but also about the remote social context of the central places of software production. This knowledge of a remote social structure becomes an important structuring tool. Knowledge of how things are done elsewhere can help bring about the same structure locally. It becomes important to look at the sources of such knowledge, and in particular at the developers' use of foreign books and web sites not just as a source of technical knowledge but also as a source of ideas for social organization.

Foreign structure, however, is not always deemed relevant to local activities. While the developers know quite a bit about how things work in remote places, they *also* know that Rio de Janeiro is no Silicon Valley. Consequently, they often consider it silly to attempt in Brazil what is known to work in California. It thus becomes important to pay attention to what outcomes can be *imagined*, and how the dubious nature of such imagination is negotiated in joint projects. Rodrigo's attempt to draw a distinction between "crazy" and "insane" dreams in chapter 8 illustrates this point. Change often comes from plans that are sufficiently "crazy" to present an ambitious step forward, yet imaginable enough to build a coalition in their pursuit.

Other Places, Other Practices

This book has relied primarily on an observation of a particular practice in a particular place. What can this account tell us about other contexts of work? I start with the question of what this book may tell us about software work in other places. I then briefly discuss whether it may be useful for understanding other kinds of work.

While additional research would be needed to examine the extent to which this perspective would fit the practice of software developers in other places, I expect that for many places such research would discover a substantial fit with the general perspective advocated here, if not with the details. This would particularly be the case for other "semi-peripheral" sites, places where the software practice has been assembled to a substantial degree but where continued work must be done to keep it up-to-date with remote standards.

My discussion of language use, for example, would be quite different if it were based on observations of software developers in one of the software capitals of Southern India. In many parts of India, English is not the language of software, but simply the language of education. It may again be somewhat different if written in Russia, where the local language is in a much stronger position vis-à-vis English than Portuguese is in Brazil, and where programming languages using non-English keywords *have* been developed. In this sense, I believe Brazil represents an intermediate position and a case worth understanding.

The notion that becoming a software developer often has more to do with learning to love the computer than pursuing lucrative employment would also not hold for India, where economic considerations appear to be the most important reason for becoming a software developer for many people. Even so, however, the larger perspective taken in this book may well apply to understanding software work in India. While Brazilian software developers learn to love software early on, but may then struggle to find "proper" jobs, software developers in India, for whom exceptional grades in high school often become a ticket to the world's largest software companies, will likely have to learn to love software after getting their jobs. (If they do not, their status in the global software world will likely remain marginal, as their foreign colleagues would see them as low-cost mercenaries rather than fellow practitioners.) Such differences fit within the broader perspective presented in this book, though understanding those two different ways of entering the world of software may be a particularly fruitful direction for a future investigation.

This book has focused on a place separated from the centers of the software world by several kinds of distance: the cost of traveling, the differences of language, national boundaries that limit the movement of both people and things, differences of government policy, national identity, local and national culture. While such different kinds of "distances" often coincide, they do not always do so. Looking at places that present specific combinations of those different kinds of distance would help refine the notions of "place" and "peripherality."

Another question concerns the extent to which the perspective taken in this chapter can be applied to other fields of endeavor. One may wonder if software is unique in the extent to which its practitioners in places such as Rio de Janeiro engage with the global world of practice, including its global culture. For many other lines of work, the local communities of practice may be all that matters to the individual practitioners. This requires a two-fold answer.

The more abstract aspects of the framework would likely be immediately applicable to a wide range of professions. Even for practices where individual practitioners rarely venture out of their local community, the substantial degree of similarity in practice points to the existence of processes that lead to synchronization of those practices between places, which likely draw on some of the same mechanisms. The perspective taken in this book would in the very least provide a starting point for analysis. For example, I have stressed that the global culture of software development provides a set of "perceptions and judgments," which include the understanding of software work as interesting and worth pursuing for the sake of intellectual stimulation. This particular way of seeing the practice is most certainly not shared by all other practices. The practitioners of a trade can instead understand their work as a matter of "service," as a matter of "honest work," as a game, or as form of political action. What is likely to be found across many different practices, however, is the pursuit of a shared understanding of the activity, whatever that understanding may be—and the ongoing struggle for a fit between this shared understanding and the material reality.

Software may well be exceptional in the extent to which the use of foreign documents by individual practitioners is important for the synchronization of the practice. Software is unique today in the abundance and accessibility of documents describing the practice. It is also unique in the extent to which such documents are *useful*. This likely has to do with the relative immateriality of the software practice. Traditional accounts of science practices, for example, commonly stress the importance of direct access to the material tools and artifacts (e.g., Collins 1974, 2001). Software developers, on the other hand, work with few physical objects apart from their computers. Their work is a disembodied, textual art. Repositories of software code and mailing lists (on which code can be shared by simply being pasted into the message) serve as virtual environments in which the objects of work reside and can be observed. Such repositories can be "visited" at little cost, as such visits do not disturb the work that occurs in them.

While there is no equivalent of this for many professions, software represents an example of a *class* of occupations that primarily involve manipulation of digital representations. Software developers simultaneously help create the technologies that enable work based on digital representations and become pioneers of such work. I expect that the number of such occupations will grow over time, and new technologies will make it possible to increasingly interleave such representations with texts. The reliance on such shareable representations may also lead to increased free sharing of elements of different practices, enabling non-software equivalents of open source production. As the analysis presented in my work suggests, however, this does not mean that place would cease to matter and may instead add power to the central sites.

What's Next for Rio's Software Developers?

As I mentioned at the end of the previous chapter, Rodrigo found in my narrative a pessimistic moral. The book had shown him, he told me, that what he wanted to do was simply not possible in Brazil and that following Alta's strategy of focusing on providing local solutions would have been wiser. Lua was perhaps an outlier, made possible by its team's privileged position at PUC and in the networks of global computer science. And as I noted in my story about Lua, its success was partly predicated on a successful separation from the local context, which could make one wonder of what use Lua can be to the place where it was born. My other stories are also full of examples of broken local alliances, as the parties made choices to build direct links with remote actors.

While my story has stressed the many difficulties of reproducing practice at the periphery and especially of peripheral innovation (which, in many ways, represents the most central form of the practice, in Lave and Wenger's terms), this focus on challenges is counterbalanced by the attention to the cumulative nature of the reproduction process. Rodrigo's project attempted to build on the Lua team's earlier success in developing an innovative programming language and, at the same time, on the broader success of the practice of software use in Brazil. The difficulty of the path pursued by Rodrigo stands out in part due to the seeming ease of the strategy pursued by Alta, which focused on building more incrementally on the already established practice. The success of Lua's team was of course enabled by the work done by the earlier generations who worked to establish the practices of computing in Brazil in the 1960s and the 1970s. *Their* work in turn built on what was done yet earlier—for example, the establishment of ITA as a part of Marshall Montenegro's plan to bring airplane making to Brazil.

I do not believe software work is likely to become fully placeless. The world of software will probably continue to be organized around a set of "meccas." Absent the collapse of the United States envisioned by Rio nerds in chapter 2, Silicon Valley is likely to remain one of those meccas for years to come. And when it is eventually eclipsed by a new center, it is unlikely that this center will be Rio de Janeiro—after all, there will be many other contenders. At the same time, however, the work of many people—those working in the decades past and those working today—helps establish the practice of software in peripheral places, giving it a seeming degree of place-lessness. It has done this by increasingly establishing compatibility between the global practice and the local context, carefully adapting the elements until they start to fit together, laying tracks upon which knowledge could travel around the world with seeming ease.

Rodrigo's own work is perhaps also best evaluated by looking at it his-torically, as yet another attempt to adapt the local context for the increas-ingly more central forms of the global software practice. We must look at the local practice that the project has fostered, rather than just at its mate-rial outcomes. Jason's discussion in chapter 8 of the way Kepler had affected him exemplifies this view of Kepler. "I met Rodrigo," Jason told me, "who is a person who showed me that it's possible to work like this. [. . .] He is there and living, right?" Web development based on Lua is yet to become widespread (and when it does, it may well happen on software written by others) and the specific configuration of resources brought together by Rodrigo has largely dissolved. I believe, however, that the effect of his work can be seen in the ways in which people who were once involved with the project find their new places in the global world of software, in their own increased ability to combine local and global resources in pursuit of their global dreams.

Notes

0   The Wrong Place

1. I use the term "software developers" to refer to people who create software and whose role in its creation requires some understanding of the inner working of this software. I also use an alternative term "programmer." I use both terms inclusively, even though within the software community a number of terms would be used depending on the situation. In general, in the United States today "programmer" is an outsider term, rarely used by the software developers themselves, who often prefer such terms as "developer" or "coder." The Portuguese cognate "programador" is used even less often by software developers in Rio where it is seen as connoting a low position in the organization. Other terms might vary from more specific job titles (e.g., "software engineer," "software architect," "systems analyst") to looser terms such as "hacker" or "software guy." Most of those terms have Portuguese equivalents, though as is the case with "programmer" the connotation sometimes varies between the two languages.

2. The San Francisco Bay Area likely accounts for between 1 and 5 percent of the world's software developers.

3. The term "worlds" also links "worlds of practice" with "social worlds," which I discuss in the next chapter.

4. When dating the beginning of free / open source software development, it is important to note that the distinction between free and proprietary software is only meaningful in the context of a particular intellectual property regime, which took its current form in the early 1980s (see Schwarz and Takhteyev 2010).

5. See Levy's (2001) account of the formation of the "hacking" culture at MIT. Also see Turner (2006) on how California's counterculture movement merged with California's computing culture.

6. In the course of my work, I found Weiss (1994) an invaluable resource on the interviewing process, Emerson, Fretz, and Shaw (1995) helped me with writing field

notes, while Becker (1998) taught me to think about the overall direction of my research.

7. Such reflexivity forms the foundation of what Burawoy (1998) calls "reflexive science."

8. I see this methodological observation as extending Giddens's (1979) argument that the actors know a good deal about the conditions of social reproduction. This is especially true when the "actors" in question are highly educated professionals. I borrow the term "ethnomethods" here from ethnomethodology (e.g., Garfinkel 1967), but I mean it the broader sense of social science—like "methods" used by the actors themselves.

9. I was particularly influenced by Marques's (2005) observation about the ambivalence inherent in the position of the peripheral practitioners, who inhabit a "contact zone" and find themselves "simultaneously copying and rejecting the models they imitate" (150). While Marques notes the resulting "impasse," however, my analysis focuses on the dynamism in "the contact zone" inhabited by the Brazilian software professionals.

10. A web development platform is a collection of software modules that serves as a foundation for building interactive web sites. Such modules normally take care of routine functionality allowing the developers to focus just on implementing features specific to their site.

11. Latour and Woolgar (1986) provide, in jest, a description of what social scientists would have to do to their subjects if they were to aim for the same degree of rigor as the biologists in the lab that they study, which includes not only full monitoring of communication but also "the right to chop off participants heads when internal examination was necessary" (256).

12. In this way, my work combined elements of traditional (though multisited) in situ ethnography with what could be seen as a case of "virtual" ethnography, drawing on many online interactions. In this sense, my fieldwork had nontrivial similarities, for example, to Nardi's (2010) study of World of Warcraft. If the "virtual" elements of ethnography are not always fully apparent in my presentation, this is because the virtual spaces in which developers collaborate with each other are normally understood by the participants as being fundamentally a part of the same reality as their face-to-face interactions. The meaning of "Rodrigo and I discussed the problem and found a solution" is understood to not depend on whether the interaction happened online or in "RL" (real life), since our identities persist between face-to-face and online interactions. (This is in contrast with World of Warcraft, where Nardi becomes "Innikka," the Night Elf when she enters the virtual world of the game.) In my presentation I try to alternate between deliberately resisting this transparency of the medium by highlighting the circumstances of the different interactions (e.g., pointing out whether they occur on beanbags or on the mailing

list), and at other times giving in to it, using phrasing that leaves the medium unspecified.

13. During my time in Brazil in 2005, I kept a journal but did not make it very detailed, because I saw my interview as my primary data. I came to regret this as I worked with my interviews in 2006 and found a need to place them in the context of my own evolving project. Consequently, when I returned to Brazil in 2007, I put a much stronger emphasis on field notes, setting myself a goal of writing around a thousand words a day and describing the events of each day in substantial detail. What I found is that this not only helped me record such events for later use but also helped focus my attention on the process of observation.

1   Global Worlds of Practice

1. An early version of this critique is provided by Contu and Willmott (2003, 2006), who laid much of the blame on Lave and Wenger themselves. Duguid (2008) provides a later analysis.

2. Unfortunately, Strauss (1979) applies the term "site" both to physical places ("mountains to climb, sites to fish," 3) and to niches in the more abstract "spaces" of activity ("sections of the sky [. . .] to examine for subspecialty purposes in astronomy," ibid.). See also Unruh's (1980) brief discussion of "geographical center(s)" of social worlds.

3. Levine's work precedes Strauss's and uses the term "worlds" more loosely.

4. Strauss warns against seeing worlds as actors, yet the primacy of individual agency is not explored in details and he frequently employs language that suggests that the worlds do act. I try to avoid this and introduce a more detailed discussion of agency and its relation to social structure later in this chapter. (My analysis can be seen as an attempt to reconcile Becker's microlevel approach to social worlds with Strauss's higher-level discussion.)

5. In this broader sense, Greek *praxis* could refer to nearly any human activity short of hard manual labor (Lobkowicz 1967).

6. The term "practice" is sometimes associated more closely with the work of Bourdieu (1977), who specifically identifies his approach as "theory of practice." Giddens (1979, 1984), on the other hand, refers to his theory as "theory of structuration." The notion of "practice," however, is quite central to Giddens's (1979) work and his use of the term is quite similar to Bourdieu's. Despite certain differences between the two approaches, their similarity is widely accepted and the two approaches are often jointly referred to as "practice theory." (Giddens 1984, which further develops the theory of structuration, no longer uses the term "practice," while also taking, in my view, somewhat of a more "micro" approach than Giddens 1979. My own work draws more closely on Giddens 1979.)

7. Giddens (1979) himself uses the term "rules." As Sewell (1992) points out, however, Giddens's use of the term "rule" is ambiguous and easily misunderstood: Giddens uses the word in its Wittgensteinian sense of "knowledge of how to proceed," not in the vernacular sense of "formal rules." From this perspective, "rules" in the everyday sense (e.g., "Employees must be at work by 9:00 a.m.") are not "rules" but "resources" in the theory of structuration.

8. While recognizing this parallel is important and fruitful, my use of the term "schema" should not be seen as a blanket endorsement of a cognitivist perspective. Attempts to analyze the internal nature of schemas can easily lead to a mechanistic view of human cognition, a road that I try to avoid. At the same time, the concept of "schema" allows us to work with units that are somewhat smaller than "culture," avoiding the temptation of imagining culture as indivisible.

9. The first three of those possibilities are analyzed by Sewell (1992). The last one is from Giddens 1979.

10. A similar type of structuration is discussed by Meyer et al. (1997), who look at the ways in which nation-states reproduce foreign models, using the term "expansive structuration" (156). (In my reading, though, the term "expansive" is used here to describe the resulting expansion of the state, rather than the expansion of the model.)

11. See also Adler 1987.

12. See, for comparison, Van Maanen and Barley's (1984) discussion of "occupational communities." In my reading of the article, Van Maanen and Barley focus primarily on the first kind of "communities" that I mention, while acknowledging the second kind. They strive to specifically move away from the broader kinds of "communities" that I discuss next.

13. This is the main reason why we should not attempt to understand them as "networks"—a term that aims to put the group in the shadow to focus on the individual ties.

14. See also Lamont and Molnár 2002 on the distinction between "symbolic" and "social" boundaries.

15. A similar approach is used by Becker (1953), on whose conceptualization I draw in chapter 3.

16. Note that Lave and Wenger (1991) use the terms "central" and "peripheral" in a nongeographic sense. "Central" forms of participation are those most significant to the community and that mark the fullest degree of membership. As I try to show, though, such central forms of participation are quite often associated with the geographic "centers" of practice.

17. My approach here attempts to bridge two traditions in sociology of work. The "cultural" approach, represented by Hughes and in a more extreme form by Becker,

stresses occupational groups as groups driven by identity. This approach, popular before the 1970s, was later criticized for politico-economic naiveté, most famously by Braverman (1974). Much of later sociology of work has, in a way, followed Braverman. While his specific pronouncements have been largely rejected (e.g., Form 1987), the literature has generally stayed focused on understanding work as a matter of labor transaction, asking how workers sell their labor, how they are controlled, and how they resist the control. The cultural approach has more recently resurfaced in literature discussing high-tech practices, especially outside sociology (e.g., Kelty 2008). Other authors who have attempted to explicitly integrate the two perspectives and have influenced my thinking on the topic included Willis (1981), Burawoy (1979), Van Maanen and Barley (1984), and Lamont (2000).

18. My notion of "moves" here is similar to the discussion of jurisdiction-shifting "moves" by professions in Abbott 1988, though I have in mind a somewhat broader range of "moves," which would include both collective and individual attempts to shift (or solidify) positions. Claiming a new mandate (perhaps by arguing that the group's culture and technique are uniquely fit for a particular role in the larger division of labor) is just one kind of move. Other examples would involve individuals aiming to use their de facto fulfillment of a role to acquire the culture and technique necessary for continued membership or, alternatively, individuals making an argument that their possession of a technique justifies creating a corresponding role in the local division of labor. I illustrate some of those moves in chapter 4.

19. Ironically, the rise of modern ICTs has dramatically helped this remote control of work. A few decades ago a manager based in the San Francisco Bay Area would have a hard time directly controlling work in India, Ireland, or Russia. Today they can. (See, e.g., Ó Riain 2000 and Aneesh 2006.)

20. I borrow the terms "center" and "periphery" from the world systems literature (e.g., Wallerstein 1974). To follow Cardoso (1972) and Evans (1979), it would be appropriate to use the term "semi-periphery" to refer to the sites that I focus on, contrasting them with the true periphery, where the practice is yet to be fully established. I stick with the term "periphery" for simplicity. It is worth noting that the term "peripheral" is also used by Lave and Wenger (1991), but of course in a very different sense: Lave and Wenger's "peripheral" participants are typically situated in the same place as the central practitioners, while practitioners working in "peripheral" sites are not necessarily novices. The two senses of the term, however, are related. Both kinds of "peripheral" participants engage in a practice over which they have less control than the more "central" members. Further, one can draw certain parallels between Evan's (1979) "dependent development" and Lave and Wenger's "legitimate peripheral participation," since in both cases a peripheral position is presented as potentially a step toward more central membership.

21. This additional value is an "externality" in that it is enjoyed by parties who do not participate in the adoption decision.

22. See Marshall's ([1890] 1927) classic analysis of the causes of industry localization.

23. In this sense, the economic notion of "network externalities" fits somewhat better with my overall framework than does Grewal's (2009) notion of "network power." Java and the English language are *resources* and as such can be a source of power for agents who can use them. Wide use of such resources makes them more powerful, not in the sense that such resources acquire their own agency but rather in the sense that such use further increases the power of agents who employ such resources at the expense of those who do not.

2   The Global Tongue

1. The quotation fixes two typos contained in the original code—one in an English word and another one in a Portuguese word.

2. My use of the term follows Grosjean's (1982). For the original usage, see Ferguson 1971.

3. Some scholars do make a claim that Brazil is diglossic between two varieties of Portuguese (e.g., Azevedo 1989; Bagno 2001).

4. To the extent that this is true for many developers but not for all of them, the result is often a "language barrier" *within* the local community of the kind that I describe later.

5. Those are the twenty-six letters that were included in the 7-bit version of ASCII (the American Standard Code for Information Interchange) in the 1960s.

6. As a global programming language, Java is designed with an assumption that the users of the software written in Java may use a variety of scripts, calendars, or sorting conventions. The *programmers*, however, are expected to use English.

7. I use the terms "middle-class," "lower-middle-class," and "upper-middle-class" as they are used by my interviewees. Roughly, lower-middle-class families are those families that can keep their kids in school through the end of high school but cannot pay for their college education or support them after high school. Occupations that are described as "lower-middle-class" usually require substantial training but offer less pay than "middle-class" occupations. "Public school teacher" is one of the most commonly cited examples of a "lower-middle-class" occupation among my interviewees. Middle-class families can support their children through college, though they cannot pay for expensive private schools like PUC.

8. A few months later, when Rodrigo set up a Portuguese mailing list for Kepler, I learned firsthand the terror of writing in a foreign language to a mailing list that archives all messages and puts them permanently in public view.

9. As I later learned, Luciano wrote most of this message by himself, and Rodrigo only corrected a few small mistakes. It also was not Luciano's first message to the list—he had sent two short messages before.

10. The result of this compilation can be found in appendix E in Takhteyev 2009.

3   Nerds from the Baixada and Other Places

1. For a somewhat different take on engineers' "love" of their work, see Kunda 1992.

2. Becker develops those ideas in the context of his analysis of "deviance" and in particular of marijuana use (Becker 1953, 1963). The framework, however, is applicable more broadly. See Takhteyev 2009 for further analysis of Becker's framework.

3. Linus Torvalds, the author of Linux, offers an eloquent explanation of what makes programming fun in chapter 5 of his book *Just for Fun* (Torvalds 2001).

4. See Petersen 1994 on the history of WordPerfect.

5. The situation has changed somewhat in recent years with the introduction of quotas for students from public schools.

6. "R$" is an abbreviation for the Brazilian real (*reais* in plural), Brazil's currency since the mid-1990s. Over the years the exchange rate between the US dollar and the real varied between one and four reais per dollar, with a 2:1 ratio in 2007. One of course must consider that salaries (and some of the prices) are substantially lower in Brazil than in the United States, so direct conversion of reais into dollars can be misleading. In 2007 the minimum monthly wage in Brazil was R$380. Monthly salaries for software developers were usually quoted around R$2,000, with R$3,000 to R$5,000 being common for "good" ones.

4   Software Brasileiro

1. Ivan da Costa Marques appears in this book in two different (though intertwined) roles. On the one hand, he was one of the key actors in the history of Brazilian informatics—a role highlighted in this chapter. In his later years, however, he turned his attention to science and technology studies, becoming not only an important source on the history of Brazilian computing, but also an important thinker on the broader issue of peripheral technology. My first encounter with Marques was through his papers, which had a substantial influence on my own thinking about Brazilian technology (see note 9 in chapter 0). Later, however, I interviewed Ivan in much the same format as my other interviewees. In cases where I do not cite specific sources, my discussion of Ivan's role is based on those interviews.

2. Cf. Latour's (1988) argument that Pasteur's bacteriology became successful on farms because farms became in essence transformed into laboratories.

3. Computing firing tables was only one of the computationally intensive tasks faced by the World War II armies. Another important one was the encryption and decryption of radio communication.

4. See Grier 1996 on the emergence of the term "program" in its contemporary meaning.

5. See Ensmenger 2010 on the development of the idea of programming as a male occupation—in particular, in relation to the testing methods used to select programmers in the 1960s. Note though, that while programming was predominantly done by men already in the 1950s, it did not become so nearly *exclusively* the male profession it is today until the 1990s.

6. Campbell-Kelly (2004, 38) reports that the RAND Corporation estimated in 1955 that there were around two hundred programmers capable of the most sophisticated development work, but probably six times as many other programmers working on simpler applications.

7. This is a very rough estimate. While most countries collect occupational statistics at a level of granularity that could be sufficient for an estimate, different countries use rather different classification systems. Consequently, aggregating the counts between different countries is difficult and is usually done only at the level of broad division by "skill level." (That is, software developers would just be counted as "professionals.") Numbers for software industry employment are easier to come by, but they provide fundamentally different counts: the software industry employs people in different occupations (e.g., accountants) and many software developers are employed by companies that fall into other sectors (e.g., banks). Additionally, sectoral statistics for developing countries that appear in print are frequently based on numbers provided by industry associations with little explanation as to how they were obtained. US government statistics for 2000 and 2006 suggested that there were likely around three million people working as "computer professionals" in the United States. I estimate that around 70 percent of them are "software developers." Brazilian statistics for the same years pointed to around 150,000 computer professionals in that country. Combined with other available statistics, this leads me to roughly estimate that the total number for the world approaches around ten million. For additional details on numbers, see appendices F and G in Takhteyev 2009.

8. See Takhteyev 2009 for several additional maps.

9. The US estimate is based on the 2000 US Census and 2006 Bureau of Labor Statistics Data. Brazilian numbers are based on the 2000 census and RAIS 2006, an occupational survey. For both countries the 2000 census gives a somewhat higher count than the 2006 occupational survey. See appendix F in Takhteyev 2009 for details, including a discussion of how "computer professionals" are counted in each country's statistics.

10. The numbers used in this section were collected in May 2008. Due to the subsequent fluctuations in the stock market, I decided not to redo the counts and used the numbers for 2008. See appendix H in Takhteyev 2009 for details, including the question of whether this number represents a fraction of the world's or just the US industry capital markets.

11. So much so that software developers occasionally refer to Microsoft by the name of the city where it is headquartered—"Redmond." Microsoft and other large companies based in the United States of course employ a substantial number of developers outside those regions. The work of such developers, however, is often focused on peripheral tasks, including what some of my interviewees call "tropicalization"—the adaptation of global software for the idiosyncrasies of the local context. To the extent that they focus on work central to the companies' products (as is sometimes the case for software developers working for major US companies in places like Bangalore), their work is directed from abroad. See Ó Riain (2000) for an example.

12. Other metrics offer a somewhat more complex though not altogether different picture. For example, see the list of companies that have contributed most changes to Linux 2.6.20 in Kroah-Hartman, Corbet, and McPherson 2009. It is important to note again that companies headquartered in the San Francisco Bay Area, the Research Triangle of North Carolina, and New York do hire developers in other places. In case of Linux in particular, the contributors include a number of developers working in Brazil. See also Takhteyev and Hilts 2010 for an investigation of the geography of open source software based on an analysis of Github.

13. It is important to note that while software platforms are predominantly developed in a small number of places, people who lead their development often come to those places from far away. In open source, the most notable examples include Linus Torvalds (the author of Linux, from Finland, now in Oregon), Guido van Rossum (the author of Python, from the Netherlands, now working for Google in Silicon Valley), and Rasmus Lerdorf (the author of PHP, from Greenland, now in Silicon Valley). Such migration toward the center has often been seen negatively, as aiding the central countries at the expense of the peripheral ones (e.g., Dedijer 1961; Johnson 1965). While the benefits that such migration has brought to the central sites seem clear, the peripheral locations may have gained from it as well in some cases. For example, Saxenian (1999, 2006) argued that the migration of engineers from Taiwan and India has helped the development of high-technology industries in such countries.

14. Marshall's ([1890] 1927) explanation of industry clustering applies remarkably well to the modern world. For a more up-to-date discussion, however, see the literature on industry clusters (e.g., Saxenian 1996; Powell et al. 2002), or the economic literature on spillover effects (e.g., Audretsch and Feldman 1996).

15. The story presented here is based primarily on existing literature, augmented with a small number of interviews with people who have worked with computers in

Brazil in the 1960s and the 1970s. I rely on the best available sources, including several works by journalists (Dantas 1988; Morais 2006), memoirs (Staa 2003), and the work of institutional historians (Freire 1993; Senra 2007).

16. Though it was IBM that suggested purchasing a computer, the company lost the bid because it could not promise to deliver a machine on the desired short schedule.

17. The next census did rely heavily on PUC participation. It should also be noted, though, its success is often also attributed in part to the more substantial assistance of USAID, which sent consultants who stayed at IBGE for months helping install and program the new equipment.

18. B205 numbers are quoted from Staa 2003. The UNIVAC price and memory are quoted from Senra 2007, while weight and energy consumption are from Weik 1961.

19. Faculty CVs are available at http://lattes.cnpq.br/.

20. This was originally pointed out to me by Sidney de Castro Oliveira who was, at the time, planning to write his doctoral dissertation on this idea.

21. According to Schoonmaker (2002), one of her interviewees, a former president of a major Brazilian computer company, described the skilled labor resources as "'eggs' left behind by 'the serpent of the market reserve'" (128).

22. According to Carvalho (2006, 96), Internet access was delayed in Brazil due to a strong commitment to OSI—a networking protocol that had been accepted by the International Standards Organization but was being supplanted by TCP/IP, the protocol used by the Internet without any official standardization process.

23. In the rest of the interview, the interviewee stresses that talking to other local practitioners continues to be important but now takes a different form—that of sharing "hints" about what to look up on the Internet. See my discussion of "pointers" in the next chapter.

24. See Marques 2007 on the negotiations surrounding Brazil's copyright laws in the 1980s.

25. For example, Linux drives popular web sites such as Google, Facebook, Wikipedia, and Yahoo! It also forms the base layer of the Android operating system for smartphones. (Apple's OSX and iOS are based on BSD, another open source operating system.)

5 Downtown Professionals

1. For a brief discussion of the Genesis incubator, see Didier, Weber, and Pimenta-Bueno 2005.

2. There is a popular Brazilian joke about an angel watching God in the process of the Creation who asks why each country is being endowed with its share of natural disasters, except for Brazil, which seems to suffer from neither blizzards nor hurricanes nor earthquakes. To this the Almighty responds: "Just wait till you see the people that I will put in that land." The joke is so well-known in Brazil that it is often shortened to "Just wait till you see the people!"

6 Porting Lua

1. For instance, Lua was ranked between fifteenth and twentieth in the TIOBE TPCI index for most of 2007 and 2008, dipping to the twenty-second position in December 2008, then rising to the twelfth position in 2011. (The index measures the popularity of programming languages by using search results.)

2. TIOBE declared C "the programming language of the year" for 2008, acknowledging the fact that C has grown in popularity in 2008, despite being one of the oldest languages in TIOBE's top twenty.

3. At the time and throughout the 1990s, Tecgraf spelled its name as "TeCGraf," highlighting the fact that the name was an acronym for "Tecnologia em Computação Gráfica" (computer graphics technology). The transition from "TeCGraf" to "Tecgraf" happened sometime between 2001 and 2003. In this chapter, I use the new spelling throughout.

4. The earliest available implementation of Lua (from July 28, 1993) contains a small number of comments in Portuguese but is otherwise written in English. DEL implementation consists of twenty-three files, twenty of which are strictly in English, including all eighteen files attributed to Luiz Henrique de Figueiredo. The other three files use a mixture of English and Portuguese for both variable names and comments. SOL implementation had minimal comments, a total of fifty-five words. All of those comments were in English, however.

5. Antônio's "many eyes" refers to Eric Raymond's (1999) pronouncement that "given enough eyeballs, all bugs are shallow"—that is, the idea that exposing the source code to other developers helps discover and fix its defects.

6. Ierusalimschy, Figueiredo, and Celes 2007.

7. The manual distributed with Lua 1.1 contains a link to "ftp.icad.puc-rio.br:/pub/lua/lua_1.0.tar.Z," which presumably represented a distribution of "Lua 1.0." The link is currently dead and it appears that its inclusion in the manual for Lua 1.1 was a mistake. A snapshot of a pre-1.1 version of Lua was later released in 2004 as "Lua 1.0" to commemorate Lua's ten-year anniversary.

8. Ten years later, Lua 5.0 (Ierusalimschy, Figueiredo, and Celes 2007) became the first scripting language to use a register-based virtual machine, which brought Lua

substantial academic interest. Additionally, in recent years, Roberto Ierusalimschy and his students have used Lua as a base for experimental work in programming language research.

9. DEL was similarly described in a Tecgraf technical report (Figueiredo 1992) and a paper presented at a conference in Brazil (Figueiredo et al. 1992).

10. The 1993 presentation was entitled "LUA: uma linguagem para customização de aplicações" (LUA: a language for customizing applications).

11. http://compilers.iecc.com/comparch/article/94-07-051.

12. Ierusalimschy, Figueiredo, and Celes (2007) describe this license as a naïve "collage and rewording of existing licenses." While the authors say that they do not remember from what sources they borrowed the text of the license, the first part of the Lua 2.1 license is identical to the license of Tcl 7.3 (released in 1993), while the rest generally corresponds to the X11 license.

13. Each of the three members of the team spent some time abroad (in different places) between 1995 and 1997, though this fact did not come up in any of my interviews. While their separation had roughly preceded the setup of the mailing list, the list did not become the locus of Lua development, even to the limited extent as that which happened with Kepler's list after Alan's departure (see chapter 8).

14. See Ierusalimschy, Figueiredo, and Celes 2007 for the text of the message.

15. A message to *lua-l*, May 10, 2011, available at http://lua-users.org/lists/lua-l/2001-05/msg00149.html.

16. Ironically, but not surprisingly, Roberto Ierusalimschy was one of the few people willing to talk extensively and on record about this conflict. Most people who were there at the time either asked to not be quoted or (more often) downplayed the complaints.

17. The discussion on the Lua list in the months leading to Lua 4.0 shows that the authors and the users were not entirely indifferent to backward compatibility, and in fact saw it as quite important. Some members also disagreed with the specific changes introduced by the new version. In the end, however, the desire to make improvement won over the concerns about backward compatibility and Lua 4.0 was received by the list with much enthusiasm.

18. In recent years, Lua has also received a number of research fellowships from Microsoft. So far, though, those have not been a major component of Lua's funding.

7 Fast and Patriotic

1. The government agency is called CAPES (a Portuguese abbreviation for "Coordination for Improvement of Higher Education Personnel") and the ranking system is known as "Qualis."

2. A message to *python-brasil*, February 2007 (my translation).

3. Though many of the people I have talked to alluded to the association between nationalism and lack of education, the author of the "fast and patriotic" comment was at the time pursuing a PhD in one of Brazil's best computer science departments. (The comment of course may have been sarcastic.)

4. Roberto was referring to our discussion of my interview with Craig, quoted earlier in this chapter.

5. As with other open source projects, the only real property involved is the trademark for the name "Lua." Others can release a modified version of Lua, but they would have to call it something else. (The Lua trademark belongs to PUC.)

6. My translation of the Portuguese text of the message. Luiz Henrique's own English translation, included in the same message, read: "Another important goal is to help spread the word about Lua to the local community and industry."

8   Dreams of a Culture Farmer

1. Lua of course was also partially funded by the Brazilian government and PUC, itself an actor tied closely to the local context. In Lua's case, however, the funding system appears to have stabilized a while back, now forming a part of the infrastructure that is almost invisible to the actors. By contrast, Kepler's alliances with the funding agencies were new and highly problematic, requiring constant renegotiation.

2. PUC earlier had an undergraduate program in "informática," but Rodrigo found it outdated.

3. See Hester, Borges, and Ierusalimschy 1997 for a discussion of CGILua in comparison to a number of the better-known alternatives. The paper does not, however, compare CGILua to PHP—a system that had been available since 1995 but was relatively unknown at the time. PHP was very similar to CGILua in a number of ways and came to dominate web development a few years later.

4. Observing the lack of collaborative relationships between industry and local research and unwilling to wait for companies to start building such relationships on their own initiative, the Brazilian government requires companies to contribute money to "sectoral funds" that are then used to fund collaborative projects such as the one described in this chapter (though typically the funding occurs on a much larger scale, according to my conversation with a FINEP grant officer). The distribution of those grants is managed by FINEP. In addition to FINEP's sectoral funds, Kepler has relied on money from other agencies, such as CNPq (an agency responsible for funding academic research) and funding agencies of the State of Rio de Janeiro. I avoid a discussion of the complex web of funding relationships, as such a description would require a tome of its own, focusing instead on FINEP, the major source of funding.

5. Favret-Saada cites the influence of Octave Mannoni on her analysis.

6. Some projects do start by scratching an itch in Rio—many if not most of my interviewees have pursued one at some point. Such projects rarely go very far, however, for lack of time and inability to find others willing to participate. It is also worth noting that many open source projects *do* proceed from the top down as funded projects. This was not the model described in the books Rodrigo was reading at the time, however.

7. My wiki was not the first web application built on top of Kepler, but it appeared to be the first that a member of the general public could easily see live and obtain the code for.

References

Abbott, A. 1988. *The System of Professions: An Essay on the Division of Expert Labor.* Chicago: The University of Chicago Press.

Adler, E. 1986. Ideological guerrillas and the quest for technological autonomy: Brazil's domestic computer industry. *International Organization* 40 (3): 673–705.

Adler, E. 1987. *The Power of Ideology: The Quest for Technological Autonomy in Argentina and Brazil.* Berkeley: University of California Press.

Anderson, B. 1991. *Imagined Communities: Reflections on the Origin and Spread of Nationalism.* Revised ed. London: Verso.

Aneesh, A. 2006. *Virtual Migration: The Programming of Globalization.* Durham, NC: Duke University Press.

Appadurai, A. 1996. *Modernity at Large: Cultural Dimensions of Globalization.* Minneapolis: University of Minnesota Press.

Audretsch, D. B., and M. P. Feldman. 1996. R&D spillovers and the geography of innovation and production. *American Economic Review* 86 (3): 630–640.

Austrian, G. D. 1982. *Herman Hollerith: Forgotten Giant of Information Processing.* New York: Columbia University Press.

Azevedo, M. 1989. Vernacular features in educated speech in Brazilian Portuguese. *Hispania* 72 (4): 862–872.

Bagno, M. 2001. Português do Brasil: herança colonial e diglossia. *Revista da FAEEBA* 10 (15): 37–47.

Barley, S. R., and G. Kunda. 2004. *Gurus, Hired Guns, and Warm Bodies: Itinerant Experts in a Knowledge Economy.* Princeton, NJ: Princeton University Press.

Bastos, M. I. 1994. *Winning the Battle to Lose the War: Brazilian Electronics Policy under US Threat of Sanctions.* Ilford. Essex, England: F. Cass.

Becker, H. S. 1953. Becoming a marihuana user. *American Journal of Sociology* 59 (3): 235–242.

Becker, H. S. 1963. *Outsiders: Studies in the Sociology of Deviance.* London: Free Press of Glencoe.

Becker, H. S. 1982. *Art Worlds.* Berkeley: University of California Press.

Becker, H. S. 1998. *Tricks of the Trade: How to Think about Your Research While You're Doing It.* Chicago: The University of Chicago Press.

Becker, H. S., and A. Pessin. 2006. A dialogue on the ideas of "world" and "field." *Sociological Forum* 21 (2): 275–286.

Botelho, A. 1999. Da utopia tecnológica aos desafios da política científica e tecnológica: O Instituto Tecnológico de Aeronáutica (1947–1967). *Revista Brasileira de Ciencias Sociais* 14 (39): 139–154.

Bourdieu, P. 1977. *Outline of a Theory of Practice.* London: Cambridge University Press.

Braverman, H. 1974. *Labor and Monopoly Capital.* London: Monthly Review Press.

Brown, J. S., and P. Duguid. 1991. Organizational learning and communities of practice: Toward a unified view of working, learning, and innovation. *Organization Science* 2 (1): 40–57.

Brown, J. S., and P. Duguid. 2000. *The Social Life of Information.* Boston, MA: Harvard Business School Press.

Brown, J. S., and P. Duguid. 2001. Knowledge and organization: A social-practice perspective. *Organization Science* 12 (2): 198–213.

Burawoy, M. 1979. *Manufacturing Consent.* Chicago: University of Chicago Press.

Burawoy, M. 1998. The extended case method. *Sociological Theory* 16 (1): 4–33.

Cairncross, F. 1997. *The Death of Distance: How the Communications Revolution Is Changing Our Lives.* Cambridge, MA: Harvard Business School Press.

Campbell-Kelly, M. 2004. *From Airline Reservations to Sonic the Hedgehog: A History of the Software Industry.* Cambridge, MA: MIT Press.

CAPES. 2009. "Documento de área—2009." http://qualis.capes.gov.br/webqualis/ ConsultaCriterio2008.faces (accessed October 6, 2011).

CAPES. 2011. "Ciência da computação." http://qualis.capes.gov.br/webqualis/ConsultaListaCompletaPeriodicos.faces (accessed October 6, 2011).

Cardoso, F. H. 1972. Dependency and development in Latin America. *New Left Review* 74:83–95.

Carvalho, M. S. 2006. *A Trajetória da Internet no Brasil: do Surgimento das Redes de Computadores à Instituição dos Mecanismos de Governança.* Master's thesis, Universidade Federal do Rio de Janeiro (UFRJ), Rio de Janeiro.

Castells, M. 2000. *The Rise of the Network Society*. 2nd ed. Oxford: Blackwell.

Ceruzzi, P. 2003. *A History of Modern Computing*. 2nd ed. Cambridge, MA: MIT Press.

Collins, H. M. 1974. The TEA set: Tacit knowledge and scientific networks. *Science Studies* 4 (2): 165–185.

Collins, H. M. 2001. Tacit knowledge, trust and the Q of sapphire. *Social Studies of Science* 31 (1): 71–85.

Contu, A., and H. Willmott. 2003. Reembedding situatedness: The importance of power relations in situated learning theory. *Organization Science* 14 (3): 283–295.

Contu, A., and H. Willmott. 2006. Studying practice: Situating *Talking About Machines*. *Organization Studies* 27 (12): 1769–1782.

Dantas, V. 1988. *Guerrilha Tecnológica: A Verdadeira História da Política Nacional de Informática*. Rio de Janeiro: LTC.

Dedijer, S. 1961. Why did Daedalus leave? *Science* 133 (3470): 2047–2052.

Figueiredo, L. H. de. 1992. "DEL: Uma linguagem para entrada de dados." TeCGraf/ICAD. http://www.tecgraf.puc-rio.br/~lhf/ftp/doc/tecgraf/del.ps.gz (accessed October 13, 2011).

Figueiredo, L. H. de, R. Ierusalimschy, and W. Celes. 1996. Lua: An extensible embedded language. *Dr. Dobb's Journal* 21 (12): 26–33.

Figueiredo, L. H. de, C. S. Souza, M. Gattass, and L. C. G. Coelho. 1992. Geração de interfaces para captura de dados sobre desenhos. In *Proceedings of SIBGRAPI '92* (Simpósio Brasileiro de Computação Gráfica e Processamento de Imagens, Águas de Lindóia, São Paulo, Brazil), 169–175.

Didier, D., E. T. Weber, and J. A. Pimenta-Bueno. 2005. Gávea angels: The birth of an angel group in Rio de Janeiro. In *Angel Investing in Latin America*, ed. E. F. O'Halloran, P. L. Rodriguez, and F. Vergara, 51–60. Charlottesville, VA: Darden Business Publishing.

Dos Santos, T. 1970. The structure of dependence. *American Economic Review* 60 (2): 235–246.

Duguid, P. 2005. "The art of knowing": Social and tacit dimensions of knowledge and the limits of the community of practice. *Information Society* 21 (2): 109–118.

Duguid, P. 2008. The community of practice then and now. In *Organizing for the Creative Economy: Community, Practice, and Capitalism*, ed. A. Amin and J. Roberts, 1–10. Oxford: Oxford University Press.

Emerson, R., R. Fretz, and L. Shaw. 1995. *Writing Ethnographic Fieldnotes*. Chicago: University of Chicago Press.

Ensmenger, N. 2010. Making programming masculine. In *Gender Codes: Why Women Are Leaving Computing*, ed. T. J. Misa, 115–141. Hoboken, NJ: Wiley & Sons.

Evans, P. 1979. *Dependent Development: The Alliance of Multinational, State, and Local Capital in Brazil*. Princeton, NJ: Princeton University Press.

Evans, P. 1995. *Embedded Autonomy: States and Industrial Transformation*. Princeton, NJ: Princeton University Press.

Favret-Saada, J. 1980. *Deadly Words: Witchcraft in the Bocage*. Cambridge: Cambridge University Press.

Ferguson, C. 1971. Diglossia. In *Language Structure and Language Use: Essays by Charles A. Ferguson*, ed. A. S. Dil, 1–26. Stanford, CA: Stanford University Press.

Florida, R. 2008. *Who Is Your City?* New York: Basic Books.

Fogel, K. 2005. *Producing Open Source Software. How to Run a Successful Free Software Project*. Sebastopol, CA: O'Reilly.

Form, W. 1987. On the degradation of skills. *Annual Review of Sociology* 13:29–47.

Frank, A. G. 1966. The development of underdevelopment. *Monthly Review* 18 (4): 17–31.

Freire, F. R. F. 1993. *Pró-Censo: Algumas Notas Sobre os Recursos para Processamento de Dados nos Recenseasmentos do Brasil. Memória Institucional–3*. Rio de Janeiro: IBGE.

Friedman, T. L. 2006. *The World Is Flat: A Brief History of the Twenty-First Century*. New York: Farrar, Straus and Giroux.

Fritz, W. B. 1996. The women of ENIAC. *IEEE Annals of the History of Computing* 18 (3): 13–28.

Garfinkel, H. 1967. *Studies in Ethnomethodology*. Englewood Cliffs, NJ: Prentice Hall.

Giddens, A. 1979. *Central Problems in Social Theory*. London: Macmillan.

Giddens, A. 1984. *The Constitution of Society: Outline of the Theory of Structuration*. Berkeley, CA: University of California Press.

Giddens, A. 1991. *The Consequences of Modernity*. Stanford, CA: Stanford University Press.

Gieryn, T. F. 1983. Boundary-work and the demarcation of science from non-science: strains and interests in professional ideologies of scientists. *American Sociological Review* 48 (6): 781–795.

Glaser, B. G., and A. L. Strauss. [1967] 1999. *The Discovery of Grounded Theory: Strategies for Qualitative Research*. New York: Aldine de Gruyter.

Graham, P. 2006. "How to Be Silicon Valley." http://www.paulgraham.com/silicon-valley.html (accessed June 19, 2008).

Grewal, D. S. 2009. *Network Power: The Social Dynamics of Globalization*. New Haven, CT: Yale University Press.

Grier, D. A. 1996. The ENIAC, the verb "to program" and the emergence of digital computers. *IEEE Annals of the History of Computing* 18 (1): 51–55.

Grosjean, F. 1982. *Life with Two Languages: An Introduction to Bilingualism*. Cambridge, MA: Harvard University Press.

Hester, A., R. Borges, and R. Ierusalimschy. 1997. CGILua: A multi-paradigmatic tool for creating dynamic WWW pages. In *SBES XI* (XI Simpósio Brasileiro de Engenharia de Software, Fortaleza, Ceará, Brazil), 347–360.

Hill, B. M. 2005. "Financing volunteer free software projects." *Advogato*. http://www.advogato.org/article/844.html (accessed March 12, 2009).

Hirschi, A. 2007. Traveling light, the Lua way. *IEEE Software* 24 (5) (September/October): 31–38.

Hughes, E. 1958. *Men and Their Work*. Glencoe, IL: Free Press.

Ierusalimschy, R., L. H. de Figueiredo, and W. Celes. 1996. Lua—An extensible extension language. *Software, Practice & Experience* 26 (6): 635–652.

Ierusalimschy, R., L. H. de Figueiredo, and W. Celes. 2007. "The evolution of Lua." In *Proceedings of ACM HOPL III (ACM SIGPLAN History of Programming Languages Conference)*. New York: Association for Computing Machinery. http://dl.acm.org/citation.cfm?id=1238844.

Johnson, H. G. 1965. The economics of the 'brain drain': The Canadian case. *Minerva* 3 (3): 299–311.

Katz, M. L., and C. Shapiro. 1986. Technology adoption in the presence of network externalities. *Journal of Political Economy* 94 (4): 822–841.

Kelty, C. 2008. *Two Bits: The Cultural Significance of Free Software*. Durham, NC: Duke University Press.

Kendall, L. 1999. Nerd nation: Images of nerds in US popular culture. *International Journal of Cultural Studies* 2 (2): 260–283.

Knorr Cetina, K. 1999. *Epistemic Cultures: How the Sciences Make Knowledge*. Cambridge, MA: Harvard University Press.

Kroah-Hartman, G., J. Corbet, and A. McPherson. 2009. "Linux kernel development: How fast it is going, who is doing it, what they are doing, and who is sponsoring it. An August 2009 update. " Linux Foundation white paper, August 2009. http://www.linuxfoundation.org/publications/whowriteslinux.pdf (accessed September 10, 2011).

Kunda, G. 1992. *Engineering Culture: Control and Commitment in a High-Tech Corporation*. Philadelphia, PA: Temple University Press.

Laird, C., and K. Soraiz. 1998. "1998: Breakthrough year for scripting." *SunWorld*. http://sunsite.uakom.sk/sunworldonline/swol-08-1998/swol-08-regex.html (accessed May 19, 2009).

Lamont, M. 2000. *The Dignity of Working Men*. Cambridge, MA: Harvard University Press.

Lamont, M., and V. Molnár. 2002. The study of boundaries in the social sciences. *Annual Review of Sociology* 28:167–195.

Latour, B. 1987. *Science in Action: How to Follow Scientists and Engineers through Society*. Cambridge, MA: Harvard University Press.

Latour, B. 1988. *The Pasteurization of France*. Cambridge, MA: Harvard University Press.

Latour, B., and S. Woolgar. 1986. *Laboratory Life: The Construction of Scientific Facts*. Princeton, NJ: Princeton University Press.

Lave, J., and E. Wenger. 1991. *Situated Learning: Legitimate Peripheral Participation*. Cambridge: Cambridge University Press.

Leontiev, A. N. [1972] 1981. The problem of activity in psychology. In *The Concept of Activity in Soviet Psychology*, ed. J. V. Wertsch, 37–71. Armonk, NY: M. E. Sharpe.

Levine, E. 1972. Chicago's art world: the influence of status interests on its social and distribution systems. *Urban Life and Culture* 1 (3): 293–322.

Levy, S. 2001. *Hackers: Heroes of the Computer Revolution*. New York: Penguin Books.

Light, J. S. 1999. When computers were women. *Technology and Culture* 40 (3): 455–483.

Lobkowicz, N. 1967. *Theory and Practice: History of a Concept from Aristotle to Marx*. Notre Dame, IL: University of Notre Dame Press.

Luzio, E. 1996. *The Microcomputer Industry in Brazil: The Case of a Protected High-Technology Industry*. Westport, CT: Praeger.

MacKenzie, D., and G. Spinardi. 1995. Tacit knowledge and the uninvention of nuclear weapons. *American Journal of Sociology* 101 (1): 44–99.

Marques, I. da C. 2000. Reserva de mercado: um mal entendido caso político-tecnológico de "sucesso" democrático e "fracasso" autoritário. *Revista de Economia da Universidade Federal de Paraná* 26 (24): 91–116.

Marques, I. da C. 2003. Minicomputadores brasileiros nos anos 1970: Uma reserva de mercado democrática em meio ao autoritarismo. *História, Ciências, Saúde—Manguinhos* 10 (2): 657–681.

Marques, I. da C. 2005. Cloning computers: From rights of possession to rights of creation. *Science as Culture* 14 (2): 139–160.

Marshall, A. [1890] 1927. Book IV: The agents of production. Chapter X. Industrial organization, continued. The concentration of specialized industries in particular localities. In *Principles of Economics: An Introductory Volume*, 267–277. London: Macmillan.

Marx, K. [1845a] 1978. Theses on Feuerbach. In *The Marx-Engels Reader*. 2nd ed., ed. R. C. Tucker: 143–145. New York: W. W. Norton and Co.

Marx, K. [1845b] 1978. German ideology: Part I. In *The Marx-Engels Reader*, 2nd ed., ed. R. C. Tucker: 146–200. New York: W. W. Norton and Co.

Meyer, J., J. Boli, G. Thomas, and F. Ramirez. 1997. World society and the nation-state. *American Journal of Sociology* 103 (1): 144–181.

Morais, F. 2006. *Montenegro: As Aventuras do Marechal que Fez uma Revolução nos Céus do Brasil*. São Paulo: Planeta.

Nardi, B. 2010. *My Life as a Night Elf Priest: An Anthropological Account of World of Warcraft*. Ann Arbor: University of Michigan Press.

Ó Riain, S. 2000. Networking for a living: Irish software developers in the global workplace. In *Global Ethnography: Forces, Connections, and Imaginations in a Postmodern World*, ed. M. Burawoy et al., 175–202. Berkeley: University of California Press.

Orr, J. 1996. *Talking About Machines: An Ethnography of a Modern Job*. Ithaca, NY: Cornell University Press.

Petersen, W. E. P. 1994. *Almost Perfect: How a Bunch of Regular Guys Built WordPerfect Corporation*. Rocklin, CA: Prima Publishing.

Polachek, H. 1997. Before the ENIAC. *IEEE Annals of the History of Computing* 19 (2): 25–30.

Polanyi, M. 1966. *The Tacit Dimension*. New York: Doubleday.

Powell, W., K. W. Koput, J. I. Bowie, and L. Smith-Doerr. 2002. The spatial clustering of science and capital: accounting for biotech firm–venture capital relationships. *Regional Studies* 36 (3): 291–305.

Raymond, E. 1999. *The Cathedral and The Bazaar*. Sebastopol, CA: O'Reilly.

Ryle, G. 1949. Knowing how and knowing that. In *The Concept of Mind*, 25–61. Chicago: The University of Chicago Press.

Samuelson, P., R. Davis, M. Kapor, and J. H. Reichman. 1994. A manifesto concerning the legal protection of computer programs. *Columbia Law Review* 94:2308–2431.

Sassen, S. [1994] 2006. *Cities in a World Economy*. 3rd ed. Thousand Oaks, CA: Pine Forge Press.

Saxenian, A. 1996. *Regional Advantage: Culture and Competition in Silicon Valley and Route 128*. Cambridge, MA: Harvard University Press.

Saxenian, A. 1999. "The Silicon Valley–Hsinchu connection: technical communities and industrial upgrading." Working Paper No. 99-10, Stanford Institute for Economic Policy Research.

Saxenian, A. 2006. *The New Argonauts: Regional Advantage in a Global Economy*. Cambridge, MA: Harvard University Press.

Schatzki, T. 1996. *Social Practices: A Wittgensteinian Approach to Human Activity and the Social*. Cambridge, UK: Cambridge University Press.

Schoonmaker, S. 2002. *High-Tech Trade Wars: U.S. Brazilian Conflict in the Global Economy*. Pittsburgh, PA: University of Pittsburgh Press.

Schoonmaker, S. 2009. Software politics in Brazil: Toward a political economy of digital inclusion. *Information Communication and Society* 12 (4): 548–565.

Schwarz, M., and Y. Takhteyev. 2010. Half a century of public software: Open source as a solution to the holdup problem. *Journal of Public Economic Theory* 12 (4): 609–639.

Senra, N. 2007. *Estatísticas Organizadas (c.1936–c.1972)*. Vol. 3, *História das Estatísticas Brasileiras*. Rio de Janeiro: IBGE.

Sewell, W. H., Jr. 1992. A theory of structure: Duality, agency, and transformation. *American Journal of Sociology* 98 (1): 1–29.

Shaw, A. 2011. Insurgent expertise: The politics of free/livre and open source software in Brazil. *Journal of Information Technology & Politics* 8 (3): 253–272.

Shibutani, T. 1955. Reference groups as perspectives. *American Journal of Sociology* 60:562–569.

Staa, A. von. 2003. Introductory notes. In *Carlos José Pereira de Lucena: Pioneiro da Informática*, ed. A. von Staa, A. L. Furtado, and S. D. J. Barbosa. Rio de Janeiro: PUC-Rio.

Strauss, A. 1978. A social world perspective. *Studies in Symbolic Interaction* 1: 119–128.

Strauss, A. 1979. "Social worlds and spatial processes: An analytic perspective." An unpublished paper. (This paper, scanned by Adele Clark, was available online in 2006, but does not appear to be available anymore.)

Strauss, A. 1982. Social worlds and legitimation processes. *Studies in Symbolic Interaction* 4: 171–190.

Takhteyev, Y. 2009. Coding Places: Uneven Globalization of Software Work in Rio de Janeiro, Brazil. PhD dissertation, University of California, Berkeley.

Takhteyev, Y., and A. Hilts. 2010. "Investigating the geography of open source software through Github." Working paper. http://takhteyev.org/papers/Takhteyev-Hilts-2010.pdf (accessed February 27, 2012).

Tigre, P. 2003. Brazil in the age of electronic commerce. *Information Society* 19 (11): 33–43.

Torvalds, L. 2001. *Just for Fun: The Story of an Accidental Revolutionary*. New York: HarperCollins.

Traweek, S. 1992. *Beamtimes and Lifetimes: The World of High Energy Physicists*. Cambridge, MA: Harvard University Press.

Turner, F. 2006. *From Counterculture to Cyberculture: Stewart Brand, the Whole Earth Network, and the Rise of Digital Utopianism*. Chicago: University of Chicago Press.

Unruh, D. 1980. The nature of social worlds. *Pacific Sociological Review* 23:271–296.

Van Maanen, J. 1988. *Tales of the Field: On Writing Ethnography*. Chicago: University of Chicago Press.

Van Maanen, J., and S. R. Barley. 1984. Occupational communities: Culture and control in organizations. *Research in Organizational Behavior* 6:287–365.

Vygotsky, L. 1978. Interaction between learning and development. In *Mind in Society*, ed. and trans. M. Cole, 79–91. Cambridge, MA: Harvard University Press.

Vygotsky, L. [1930] 2002. Орудие и знак в развитии ребёнка. In Психология. Moscow: EKSMO-Press.

Wallerstein, E. 1974. Dependence in an interdependent world: The limited possibilities of transformation within the capitalist world economy. *African Studies Review* 17 (1): 1–26.

Weiss, R. S. 1994. *Learning from Strangers: The Art and Method of Qualitative Interview Studies*. New York: Free Press.

Weik, M. 1961. "A third survey of domestic electronic digital computing systems." Ballistic Research Laboratories, Report No. 1115. http://ed-thelen.org/comp-hist/BRL61.html (accessed October 14, 2011).

Wellman, B., and K. Hampton. 1999. Living networked on and offline. *Contemporary Sociology* 28 (6): 648–654.

Willis, P. 1981. *Learning to Labor: How the Working Class Kids Get Working Class Jobs*. Morningside ed. New York: Columbia University Press.

Xiang Biao. 2006. *Global "Body Shopping": An Indian Labor System in the Information Technology Industry*. Princeton, NJ: Princeton University Press.

Zook, M. A. 2002. Grounded capital: Venture financing and the geography of the Internet industry, 1994–2000. *Journal of Economic Geography* 2:151–177.

Index